Climate Change and Social Movements

Climate Change and Social Movements

Civil Society and the Development of National Climate Change Policy

Eugene Nulman
Lecturer, Birmingham City University, UK

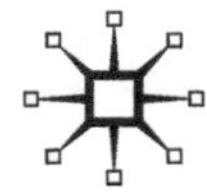

First published 2015 by
PALGRAVE MACMILLAN

Palgrave Macmillan in the UK is an imprint of Macmillan Publishers Limited, registered in England, company number 785998, of Houndmills, Basingstoke, Hampshire, RG21 6XS.

Palgrave Macmillan in the US is a division of St Martin's Press LLC, 175 Fifth Avenue, New York, NY 10010.

Palgrave is the global academic imprint of the above companies and has companies and representatives throughout the world.

Palgrave® and Macmillan® are registered trademarks in the United States, the United Kingdom, Europe and other countries.

ISBN: 978–1–137–46878–9

This book is printed on paper suitable for recycling and made from fully managed and sustained forest sources. Logging, pulping and manufacturing processes are expected to conform to the environmental regulations of the country of origin.

A catalogue record for this book is available from the British Library.

A catalog record for this book is available from the Library of Congress.

To my parents, Vladimir and Florida

Contents

List of Tables and Figures

Tables

Figures

List of Abbreviations and Acronyms

AOSIS	Alliance of Small Island States
BAA	British Airport Authority
BIS	Department of Business, Innovation, and Skills
CaCC	Campaign against Climate Change
CAN	Climate Action Network
CBI	Confederation of British Industry
CCC	Committee on Climate Change
CJN	Climate Justice Network
CO_2	carbon dioxide
COP	Conference of the Parties
DECC	Department of Energy and Climate Change
DfT	Department for Transport
EAC	Environmental Audit Committee
EC	European Community
EDM	early day motion
ENGO	environmental non-governmental organization
ERRB	Enterprise and Regulatory Reform Bill
EU	European Union
FCCC	Framework Convention on Climate Change
FoE	Friends of the Earth
GIB	Green Investment Bank
HACAN	Heathrow Association for the Control of Aircraft Noise
INC	Intergovernmental Negotiating Committee
IPCC	Intergovernmental Panel on Climate Change
ISA	individual savings account
MP	Member of Parliament
NGO	non-governmental organization
NIMBY	Not In My Backyard
NoTRAG	No Third Runway Action Group
ONS	Office of National Statistics
POS	political opportunity structures
PSNB	public sector net borrowing
PSND	public sector net debt
Quango	Quasi-autonomous non-governmental organization

REDD	Reducing Emissions from Deforestation and Forest Degradation
UK	United Kingdom
UNCED	United Nations Conference on Environment and Development
UNEP	United Nations Environment Programme
UNFCCC	United Nations Framework Convention on Climate Change
US	United States of America
WCC-2	Second World Climate Conference
WMO	World Meteorological Organization
WWF	World Wide Fund for Nature

Acknowledgements

I am indebted to a number of people who have helped me complete this research, from its early stages as a doctorate thesis to this final stage. First and foremost I would like to thank Chris Rootes for all of his assistance and inspiration as my supervisor during my masters and PhD studies. Chris has been there throughout all the stages of the process by introducing me to social movement theory, helping to develop my research competencies, and providing me with continuous advice. Chris has always provided me with the support I needed to complete this project by being both meticulous in editing and scrutinizing my work and allowing me the space to develop my ideas and make my thesis my own. Chris has also been helpful in examining some of the many changes I have made to the thesis for this book and in assisting me in finding employment, without which I may have given up on the book.

I must also thank other colleagues for their helpful comments on the thesis, book, and specific chapters, including Clare Saunders, Neil Carter, Ben Seyd, Loraine Bussard, and Alex Norman. I would like to thank many others including anonymous peer reviewers for the book and related publications, as well as those who provided constructive criticisms during conference presentations on the subject. I should also use this space to thank the various campaigners, policymakers, and business representatives that took time away from their busy schedules to speak to me.

I would like thank my family for offering their help and support whenever I needed it and my friends for dragging me away from my work to ensure I enjoyed my experience and was not holed up in my office with only my books to keep me company. Last, but certainly not least, I would like to thank my partner, Francesca Comparone, who was always there for me despite living in another country and whose love has always sustained me.

Foreword

Climate change is widely, if not universally, acknowledged to be the greatest challenge that we face. It is not surprising, then, that it should be a prominent object of public policy, nor that environmental movements and NGOs should have mobilized to make it and keep it so. Yet it would be wrong to assume that policy on climate change is a simple product of pressure by organized environmental activists.

Climate change is unusual among environmental issues in that it was not placed on the agenda by environmental movements or NGOs but by scientists, who by the 1970s had begun to notice a sustained rise in average global surface temperatures and were concerned about its possible impacts upon global weather patterns. The World Meteorological Organization, a specialized agency of the United Nations, served as a conduit by which scientists' concerns might be communicated to national governments, and in 1988 an Intergovernmental Panel on Climate Change (IPCC) was set up to evaluate the scientific evidence on climate change. This, in due course, led to the United Nations Framework Convention on Climate Change (UNFCCC) in 1992 and thence to the Kyoto Protocol.

If climate change is a 'super-wicked' issue for governments, it is also a difficult issue for environmental movements and NGOs. Environmental NGOs and campaigns have mostly attracted attention, and achieved results, by focusing upon particular issues neglected by governments and by seeking specific remedies to the particular problems they identified. They were not accustomed to demanding, or seeking to mobilize the public to demand, systemic change. For that reason, they were often denounced as 'merely reformist', especially by the dwindling band of devout socialists and other leftists for whom capitalism is the root of all evil, and its abolition or transcendence the necessary condition of any meaningful assault on our environmental predicament.

Environmental NGOs' rationale was to seek an endless stream of 'small wins' that might improve environmental conditions for humans and other animals and at the same time sustain the interest and optimism of their mass constituencies, and thus to maintain the momentum of an environmental movement. Climate change, however, was too big and complex for that. Nobody – and certainly not environmental

NGOs, whose veterans were, from bitter experience, acutely aware of the difficulties of educating the public about less visible environmental issues – could have much confidence that the public would understand or, having understood, would prioritize action on a locally 'abstract', even hypothetical problem whose most immediate and most severe effects would be felt far away geographically and at some indeterminate time in the future. Climate change was, is, a *global* issue at a time when there was, and perhaps still is, no genuinely global environmental movement and when global environmental consciousness – in the sense of a proper understanding of the ecological interdependence and interpenetration of the local parts with the global whole – remains elusive even for the minority of people who are educated about such matters.

Effectively addressing climate change requires action at a level and in an arena – the international – in which environmental NGOs were ill-resourced to compete, either with national governments or increasingly powerful transnational corporations. Recognizing that, environmental NGOs from Europe and North America cooperated to form Climate Action Network, a specialized environmental network, designed to lobby at the appropriate transnational level.

The limitations of outsourcing lobbying on climate change to Climate Action Network became evident with the failure of the 2000 Conference of the Parties (COP) to the UNFCCC in The Hague, when US proposals to include nuclear energy in the Clean Development Mechanism incited anti-nuclear European environmental NGOs such as Friends of the Earth and Greenpeace to take up an issue made more urgent by the failure of the COP to reach an agreement and the subsequent withdrawal of the US, then the principal emitter of climate-jeopardizing greenhouse gases, from the Kyoto process.

Some national governments responded to scientists' alarums about climate change more readily than others. If on some environmental issues, urban air pollution notable among them, the UK was a laggard among long-industrialized countries, UK governments were leaders on climate change and the need for action to mitigate it by reducing dependence on the most carbon-intensive fossil fuel, coal. It could not escape notice that there was an element of special pleading in this, because the UK was newly endowed with abundant North Sea gas and the Conservative government of Margaret Thatcher was determined to break the power of the coal miners' union that was blamed for bringing down the Heath Conservative government in the 1970s. Nevertheless, there is no reason to doubt that Thatcher, the first science graduate to head a UK government, was sincere in her belief, publicly articulated as early as the 1979 meeting of the G7, that the burning of fossil fuels was endangering the balance of the global climate.

Although climate change quickly became an issue of cross-party consensus in the UK, it was during the Prime Ministership of Tony Blair (1997–2007) that the UK took up climate change in international forums with missionary zeal. Yet, when it came to domestic action to mitigate climate change, the Blair government timidly demanded evidence of public support for action before it would itself act. Labour ministers professed concern about climate change, but insisted that the environmental movement should mobilise the public in order to give the government confidence that, if and when it acted, it was not getting ahead of public opinion.

The environmental movement was urged to follow the example of Make Poverty History, which, with explicit support from Labour ministers, succeeded in mobilizing large numbers of people to demand fairer trade, better aid and debt relief for the poorest developing countries. By the time Stop Climate Chaos, a campaign umbrella for a wide range of aid, trade and development as well as environmental NGOs, was launched in 2005, Friends of the Earth was already campaigning for a binding legislative commitment to reducing the UK's greenhouse gas emissions. The Climate Change Bill quickly gained the support of most Members of Parliament and in due course, the Labour government adopted it. Characteristically, when it adopted the Bill, the Labour government justified doing so by its aspirations to global leadership on the issue. The result was a world-leading piece of legislation, and a much vaunted victory for the environmental movement.

The environmental movement's elevation of climate change as an issue demanding urgent action rather than merely fine words not only helped secure passage of the Climate Change Act and smoothed the path for the Green Investment Bank, but it provided a universalizing frame for the campaign against the proposed third runway at Heathrow airport, which might otherwise have remained a local issue.

Not every environmental movement demand finds a receptive audience or a smooth path to legislation, regulation and / or implementation. For that reason, we need research that illuminates the conditions under which environmental movements may succeed in shaping policy and that identifies the mechanisms by which they do. Eugene Nulman's book has the great virtue of doing just that by examining these three quite different but highly salient cases of environmental policymaking in recent British history.

Every issue and every campaign has characteristics and specificities peculiar to it, and political opportunities vary from polity to polity and time to time. By comparing three nearly contemporaneous climate-related campaigns in a single major polity, Eugene Nulman has in this book succeeded in illuminating the mechanisms and strategies by which policies were formed and forced onto the national political agenda. Thus he succeeds in addressing that most elusive of topics in the social scientific study of social movements – the impact of public mobilization and NGO campaigns upon public policy.

However, the race is not yet run, and it remains possible that environmental movements will yet have a major role to play, not only in shaping policy, but in ensuring the effective implementation of policies designed to address the looming threat of climate change. Certainly, the conditions of political competition that provide the opportunities and the 'policy window' for action on climate change have changed since the election of a majority Conservative government, the decimation of the Liberal Democrats, the most environment-friendly of the major UK parties, and the introversion of the Labour Party as it struggles to find a new leader and a convincing rationale.

Although the Conservative government remains committed to international action to mitigate climate change, it remains to be seen whether its domestic action on climate change will surmount the constraints that stem from the government's fiscal conservatism, its ambitions for economic growth, and its imposition of increasingly severe austerity upon public sector agencies and local government. If it does not, environmental activists may be given new grounds for mobilization. And then some new permutation of the mechanisms and strategies that Eugene Nulman so ably discusses will come into play in determining the impacts of renewed movement action.

Christopher Rootes
Harbledown
June 2015

1
Introduction

Flooding of densely populated urban areas (McGranahan *et al*, 2007); droughts leading to widespread famine (IPCC, 2014, ch. 9); mass extinction of plant and animal species (Thomas *et al*, 2004); the devastation of entire nations (Barnett and Adger, 2003). This is the harrowing picture painted by the scientific community as they attempt to predict the effects of climate change. Some of these effects are already being experienced as the rates of abnormal and extreme weather events connected with climate change rise (Coumou and Rahmstorf, 2012).

Climate change, as we experience it today, is largely a product of the intense increase of greenhouse gas emissions since the period of industrialization. It is a global problem that is still in search of a solution. International negotiations have failed to adequately address the major task of cutting greenhouse gas emissions. Political calculations made by policymakers thus far, given the present national, international, and economic conditions, have not added up to produce policies that will significantly reduce greenhouse gas emissions. What can change this calculation, generating enough pressure to force serious mitigation measures to be taken?

It is possible that the devastation caused by additional warming will force the hand of politicians, but by then it may take decades to reverse the warming trend (Schlanger, 2014). Another option is pressure from civil society, from social movements. Activists and organizations can work to pressure policymakers to address climate change. They attempted this when international negotiations on mitigation began to take place, pushing for a strong global treaty. Chapter 2 describes this process and the influence the climate change movement had on international environmental policy. Despite these efforts, and some small successes, strong international policy has failed to materialize. In large part, this is due to national-level calculations by major greenhouse-gas-emitting countries. For the international community to work together on climate change, national policymakers will need to redo their calculations based on new information. Social movements can play a role by influencing these national-level calculations in a wide variety of ways, including the development of important national policies to reduce emissions and help mitigate climate change. This book explores the outcomes of activists and organizations that make up the climate change movement in their attempts to influence national climate change policy following what the movement perceived as a failure of international policymaking. Specifically, the book looks at the changes the climate change movement in the UK was able to make regarding three important national policies concerning emission targets, carbon-intensive infrastructure, and funding for renewable energy.

The climate change movement and movement outcomes

The climate change movement is an amalgam of loosely networked individuals, groups, and organizations springing out of the environmentalist, development, anti-capitalist, and indigenous movements, combined with a new wave of activists and groups that had no previous ties to other social movements. A **social movement** is 'a loose, noninstitutionalized network of informal interactions that may include, as well as individuals and groups who have no organizational affiliation, organizations of varying degrees of formality, that are engaged in collective action motivated by shared identity or concern' (Rootes, 2007a, 610).

Like other movements, the climate change movement has a general objective – in this case, the objective is to reduce the negative impacts of climate change on people and the planet. The successes the movement has in achieving this goal are known as **outcomes.** Outcomes represent conscious attempts at influencing political, economic, and social change. They can be contrasted with **unintended outcomes** or consequences of movements. It must be said that outcomes and consequences are not always easy to distinguish. Different individuals, **social movement groups**, and **social movement organizations** may have different aims and objectives. However, by defining the climate change movement as having the goal of 'reducing the negative impacts of climate change on people and the planet', only those efforts will be considered here.

One type of outcome sought by many movements, including the climate change movement, regards government policy (Amenta and Caren, 2007). Movements can attempt to achieve all or some of their goals by getting policymakers to use the power and resources of government to address particular issues.

The three campaigns

This book examines the climate change movement's attempts to influence national climate change policies in the United Kingdom. The United Kingdom represents an interesting context for the exploration of climate change movement outcomes for a variety of reasons. First, the UK has been an outspoken leader of international negotiations on climate change since the premiership of Tony Blair but has failed to implement effective domestic policies prior to the cases examined here. Second, the UK was on track to meet its climate change targets as set in the Kyoto Protocol as a result of unrelated policies that led to the reduction of coal use. This meant that the UK had less incentive to implement climate change policies nationally. Third, the climate change movement had campaigned for several national-level mitigation policies, providing a variety of cases to examine. I examine three of these **campaigns** focused on three separate challenges to reducing greenhouse gases: legislating emissions targets, opposing new carbon-intensive infrastructure, and increasing funding for renewable energy.

1. Friends of the Earth's *Big Ask campaign* called on the government to legislate an annual 3 percent cut of greenhouse gas emissions that would result in a reduction of 80 percent by 2050. Friends of the Earth, engaging its local branches and celebrity supporters, mobilized the public to lobby the government, getting nearly 200,000 people to participate in the campaign. While eventually succeeding to pressure the government to propose their own climate change bill, the government's draft was a weaker bill than campaigners had in mind. The campaign reignited to call for amendments to the government's bill.
2. The *campaign against a proposed third runway at London's Heathrow Airport* represents a campaign to stop the construction of carbon-intensive infrastructure, or as a spokesperson for the group Campaign against Climate Change (CaCC) put it: 'This is not just about Heathrow, this is about drawing a line in the sand against big investment decisions that are locking us into a headlong plummet into climate catastrophe' (in Vidal, 2008). The expansion of Heathrow Airport would result in increased flights, and research at the time found that growing aviation emissions could hamper attempts to mitigate climate change. Local campaigners mobilized around various local issues including air and noise pollution as well as the demolition of houses and other buildings. At the same time, a key figure in the campaign also encouraged climate change activists to mobilize against the airport expansion plans, which resulted in sustained **direct action** that generated

unprecedented levels of media attention. Campaigners were able to convince the major opposition political party to oppose the third runway, which resulted in political jousting during the election cycle. When two opposition parties formed a coalition government in 2010, they put an end to the threat of the third runway at Heathrow...for the time being.

3 The book discusses the case of government investment into renewable energy, energy efficiency measures, and low-carbon technologies. Research had shown that to meet the UK's climate change targets, over £160 billion was needed by 2020 just for the centralized energy infrastructure. This meant large levels of private-sector investments were needed, but the private sector was hesitant to finance high-risk projects without government support. In order to reduce the risk and increase investment, environmental non-governmental organizations (ENGOs) developed the idea of and campaigned for a *green investment bank*, a legislated institution that would provide public investment for climate change mitigating projects such as energy efficiency measures and renewable energy sources. A closely networked coalition formed in order to lobby the government, provide policymakers with detailed policy information, and mobilize the support of the finance sector. In doing so, they attracted the attention of policymakers who adopted their idea but proposed an institution that was very weak. The campaign then worked to strengthen the policy as it made its way through Parliament, but it failed to realize many of the coalition's desired amendments.

Chapter 3 discusses the histories of these campaigns in order to gain a better understanding of the efforts involved. The chapters that follow will examine the cases regarding the central research questions that surround movement outcomes: *what, when,* and *how.*

Research questions: what, when, and how

Many questions arise when researching social movement outcomes. Naturally, when exploring a given case, we are interested to know if any outcomes were achieved by the movement, and if so, which. This is the *what* question of social movement outcomes (that is, *what* are the outcomes achieved by the movement?). Early social movement scholars have answered the *what* question relatively simply: success or failure. While success and failure can be attributed to various types of outcomes, such as 'new advantages' for the beneficiaries of a social movement or 'acceptance' of a movement or organization by powerholders (Gamson,

1975), the problem arises as to what constitutes success or failure. If a movement is composed of a variety of organizations, groups, and individuals, who determines the precise accomplishment that would indicate success or failure (see Giugni, 2004)? In trying to circumvent this binary position, some scholars attempted to use a scale of success and failure and apply it to more 'objective' criteria. For example, Giugni's important text on the subject of movement outcomes examines the environmental, anti-nuclear, and anti-war movements' influence on national and local environmental spending, nuclear energy production and the number of construction permits granted for new nuclear energy plants, and 'defense' spending, respectively (Giugni, 2004). Other studies have also utilized similar quantitative measures (Agnone, 2007; Andrews, 1997; Giugni and Yamasaki, 2009; Kolb, 2007; McCammon *et al*, 2001; McAdam and Su, 2002). These quantitative measures allow us to see the extent to which movements influence these variables beyond a dichotomy of success and failure, but they mask the nuance of demands social movements make, and they fail to test outcomes on other important factors concerning policy.

In order to answer the *what* question as thoroughly as possible, we must break down the policy process into component parts and examine outcomes as a scale in each component, since it is possible that one component was affected by the movement, while another was not. In order to do this, I have designed the Policy Outcomes Model, which includes the following components: policy consideration (the extent to which pro-movement policies are considered), political support (the extent to which policymakers ally themselves with pro-movement policies), political action (the extent to which political action is taken to deliver pro-movement policies), desired change (the extent to which policy change is formulated and functions to serve movement goals), and desired outcome (the extent to which the policies achieve a movement's broader goals). Chapter 4 shows the extent to which the campaigns were able to influence each of these components.

The next major question that social movement outcomes pose is *when* do movements have the ability to influence policy? This question pertains to the political and social contexts in which movements operate. Much of the outcomes literature has observed that movements were able to influence policy when 'the time...was right' (Ganz, 2004, 178). This is particularly important to the political process school of social movement research that sees political opportunities and political opportunity structures as crucial to movement outcomes. The general argument is this: variables outside of the control of movements, and largely within the political arena, will determine the ability of movements to engage with policymakers and create policy change. These variables can be structural, in that they represent long-term conditions within political institutions, or dynamic, in that they are readily subject to change. For example, political systems that are 'open' to movements by having citizen referenda or high levels of elite conflict may result in movement outcomes, whereas closed systems may result in fewer outcomes (Kitschelt, 1986; Kolb, 2007). This line of argument is put to the test in Chapter 5. The chapter pays special attention to the opening of policy windows as an answer to the *when* question.

The final question I will tackle in this book is *how* are outcomes achieved? I will do so by separately examining two key concepts found in the social movement outcomes literature: strategic leadership and mechanisms.

Strategy, in some regards, can be viewed as the antithesis of the political process school. Whereas the *when* question looks at variables outside of the control of movements, strategy focuses particularly on the agency of movement actors in devising plans to influence policy. Strategies are often a product of leadership decisions made with regards to a particular campaign or a long-term group or organizational decision. These can be understood as strategic domains. Within these domains, several strategic decisions can be made pertaining to how open the campaign is to others (extension decisions); what relationship the campaigning organizations have with external actors, target institutions, and between themselves (relational decisions); and what tools and tactics will be used to attempt to achieve their goals (tactical decisions). I explore strategy and the role of strategic leadership in Chapter 6.

The second part of the *how* question concerns mechanisms, or the causal processes by which social movements can influence policy change. When looking to see how outcomes were achieved, it is important to examine all possible mechanisms in order to consider the relative importance of each without missing a crucial causal pathway to policy change. However, such thorough examinations have rarely been done (although see Kolb, 2007). In his 2007 book on social movement outcomes, Kolb produces a list of five mechanisms (disruption, public preference, political access, international politics, and the judicial mechanism), which I will test. *Disruption* regards the ability of movements to interrupt the normal processes of political or social life to the point that policymakers concede to movement demands in order to return to a state of normalcy. *Public preference* refers to a movement's ability to influence public opinion and issue salience to the point that policymakers make pro-movement policy decisions due to electoral concerns. The *political access mechanism* posits that movements can influence policy by gaining new political rights or powers for themselves or their beneficiaries. *International politics* refers to several means by which movements can leverage others outside of the national political arena in which one is campaigning. The *judicial mechanism* is a process by which movements can influence policy by referring their complaints to the judicial system and for those judicial decisions to then change policy. I examine each of these mechanisms in Chapter 7 before concluding (Chapter 8).

This book seeks to contribute to our understanding of social movement outcomes by taking a detailed look at several cases within the same structural context but with differing outcomes that are a product of campaigning by different groups and individuals within the same movement. This allows us not only to gain a better understanding of the role of civil society in the formation of national policies, but also offers lessons to campaigners on how such policies can be made, shaped, brought to light, and legislated. By specifically looking at the climate change movement, this book also allows us to better understand the processes involved in attempts to mitigate the devastating consequences of climate change.

2
Brief History of Climate Change Policy and Activism

Climate change mitigation requires an international effort. But as I will argue in this chapter, policies at the national level are an important element of international progress on climate change. Environmentalists and activists that form the climate change movement did not initially focus on national mitigation policies. Their efforts arose from a context of international negotiations that developed as scientific data on the subject increased. The climate change movement worked to influence these international negotiations, but they failed to have a significant impact as key developed countries' national interests did not align with a strong climate treaty. Europe, however, was seen as an important force for pushing negotiations forward, and the United Kingdom in particular was looked up to as an important actor. This started with the premiership of Margaret Thatcher.

Margaret Thatcher and climate change

On 27 September 1988, Margaret Thatcher addressed the UK's national academy of science, the Royal Society. Although Thatcher had a scientific background, having completed her undergraduate degree in chemistry at Oxford and working as an industrial chemist for four years (Agar, 2011), she spoke to the Royal Society as the prime minister of the UK. While opinion is divided on her legacy, that Tuesday in September, the Conservative prime minister's speech set a precedent for an issue that nearly 30 years later has still not been adequately addressed by world leaders. Between speaking about the importance of science and raising concerns about ozone depletion and acid deposition, Thatcher spoke to the group of esteemed scientists about the growing evidence of the rise in greenhouse gases 'creating a global heat trap which could lead to climatic instability. We are told that a warming effect of 1°C per decade would greatly exceed the capacity of our natural habitat to cope. Such warming could cause accelerated melting of glacial ice and a consequent increase in the sea level of several feet over the next century' (Thatcher, 1988). She expressed concerns for the Maldives and noted the record temperatures in the 1980s. Weeks later she repeated these concerns in her speech to the Conservative Party Conference. These are widely considered her first public remarks on the issue of climate change (see for example, Hulme, 2013, 9) and the first by any major world leader.

While she may have been the first world leader to address the issue, she was certainly not the first person to discuss it within the political arena. In 1983, the United States Environmental Protection Agency issued a report on climate change entitled 'Can We Delay a Greenhouse Warming?' (Kutney, 2014, 138). Just several months prior to Thatcher's speech to the Society, the United States Senate Committee on Energy and Natural Resources was held. It was here that climatologist and director of NASA's Goddard Institute of Space Studies James Hansen famously stated that he felt '99-percent confident' that an observed high temperature trend was in fact a 'real warming trend' rather than natural variability (Christianson, 1999, 196–8).[1] The fact that the summer of 1988 was sweltering was said to give added salience to the issue.

Indeed, international climate change policy was already being discussed by high-ranking policymakers. In June 1988, delegates from nearly 50 countries attended the Canadian government-sponsored World Conference on the Changing Atmosphere held in Toronto, which called for increased resources into research and monitoring of the global climate and establishing an international framework for dealing with the issue (Soroos, 2002, 126; Zillman, 2009).[2] That same year, the UN General Assembly passed Resolution 43/53 stating that climate change was a 'common concern of mankind' (Fitzmaurice and Elias, 2005, 339, n.78), and the first meeting of the Intergovernmental Panel on Climate Change (IPCC), which grew out of joint efforts by the World Meteorological Organization (WMO) and the United Nations Environment Programme (UNEP) (Bolin, 2007, 47), was held. Concerns about climate change were also growing beyond scientific circles. Public opinion polls found that climate change became an issue of concern for a majority of adults in various developed countries (Weart, 2004, 116–7), and by January 1989, climate change was such a mainstream issue that *TIME* magazine reworked their popular 'person of the year' issue to 'planet of the year'.[3]

Although the issue of climate change was growing increasingly salient, Thatcher stunned many British commentators with her speeches in 1988. She proceeded to stun the international community during her well-known mention of climate change in her speech at the United Nations General Assembly in November 1989, stating that '[t]he most pressing task which faces us at the international level is to negotiate a framework convention on climate change' (Thatcher, 1989). Thatcher was indeed a pioneer in speaking about the issue, but her efforts started some ten years prior.

In 1979, just one month after taking office, Margaret Thatcher met with the heads of state of France, West Germany, Italy, Japan, the United States, Canada, and the European Commission at the fifth G7 Summit in Tokyo. There, prior to the official summit, she was interviewed by BBC journalist Bob Friend. In her interview she stated that '[we] should also be worried about the effect of constantly burning more coal and oil because that can create a band of carbon dioxide round the earth which could itself have very damaging ecological effects' (in Moore, 2013, 448). Her concerns were acknowledged by the G7 and included in their declaration, reading: 'We need to expand alternative sources of energy, especially those which will help to prevent further pollution, particularly increases of carbon dioxide and sulfur oxides in the atmosphere' (Declaration: Tokyo Summit Conference, 1979, 5).

However, this was only a minor note in a document largely concerned with the price of oil. The statement on carbon dioxide followed a paragraph of the declaration pledging G7 countries would 'increase as far as possible coal use, production, and trade' (Declaration: Tokyo Summit Conference, 1979, 5).

Thatcher would not take this pledge too seriously in the years that followed, knowing that increasing coal production would work to the benefit of the miners' union, a major player in bringing down the 1970–74 Conservative government. By 1984, Thatcher's push to close uneconomic coal pits resulted in a clash with militant coal miners, a clash that ended with the closure of more mines than originally intended (see Blundell, 2008, ch. 15). Electricity markets began to shift from coal to natural gas supplied from the North Sea. The switch to gas, also buoyed by Thatcher's efforts to privatize energy companies, had the unintended consequence of lowering the UK's greenhouse gas emissions. Thatcher's motivations for changing policy were ideological, political, and economic, but not environmental.

In fact, Thatcher's environmental record was fairly poor. While she was prime minister, Britain became known as 'the dirty man of Europe'. Although the exact origin of this epithet is unknown (Porritt, 1989), it was connected with the UK's aversion to European environmental regulations during Thatcher's tenure, particularly regarding the 1988

Council of the European Communities Directive that called for targets to reduce annual emissions of sulfur dioxide and nitrogen oxides from large combustion plants. The UK became 'the main stumbling block' to negotiations, stymying talks for six years before the legislation was secured (Jones, 1985). Under Thatcher's leadership, the Conservative government went from being 'virtually on top' in implementing and complying with the European Communities' environmental directives to 'seeking to disregard or subtly circumvent the standards laid down in certain Directives' (Porritt, 1989, 493), including the levels of nitrates in water supplies. Thatcher's government also worked to undermine international attempts to reduce chlorofluorocarbons in an attempt to address ozone depletion, pushing for low reduction targets and hesitating to ask industry to label products or phase out ozone-depleting chemicals (Porritt, 1989, 493).

While the UK was weary of European Community (EC) policy, it felt more comfortable in international policymaking, because it was better able to preserve sovereignty, the US could help protect it from the influence of the EC, and the passage and implementation of costly policies would be stalled by lengthy negotiations (Cass, 2006, 24–5).[4] This meant that while the UK became a prominent figure within international efforts to mitigate climate change, it was largely as a means of pushing for a quick convention to lay out 'good climatic behavior' while eschewing concrete commitments to greenhouse gas reductions (British representative to the United Nations, Sir Crispin Tickell, in Pettenger, 2007, 41). In 1989, while EC countries called on negotiations to begin at the UNEP council meeting in Nairobi and the US felt negotiations to be premature, the UK drafted a compromise that, in effect, delayed negotiations until the 1990 IPCC report (Cass, 2006, 25–6). Although the UK went on to push for binding international agreements on efficiency and forestry regarding climate change, their domestic policies at that time included a large road-building scheme and cutting energy-efficiency funding (Cass, 2006, 26).

The short vignette above about Margaret Thatcher's role in climate change policy demonstrates that while rhetoric toward the environment changed, policy positions had not. Environmentalists were not unaware of this and criticized Thatcher's empty words. Friends of the Earth England Wales Northern Ireland (FoE) director Jonathon Porritt (1989, 491–2), instead, saw an opportunity:

> Some commentators (including spokespersons in both the Labour Party and the Democrats) have instantly dismissed her intervention

> as a wholly cynical exercise in vote-catching, a massive rhetorical impertinence when set against the backdrop of environmental policy-making in this country since 1979. ...What is most important, however, is that Mrs. Thatcher is now 'on the record' about the environment in a way which no one in the Green Movement had imagined possible before. The rhetorical declarations are important, not just because they are totally unprecedented, but because they constitute a serious challenge to Ministers, MPs and Councillors within the Conservative Party to live up to them (Porritt, 1989, 491–2).

At the very least, Thatcher's statements provided the environmental movement with an opportunity to discuss the issues publicly and have the ear of the media (Carvalho, 2000). The movement attempted to exploit these opportunities in addressing the international climate change negotiations that began to take place at around that time.

Early environmental movement activity on international climate change negotiations

It was not until the late 1980s that the environmental movement, and what later developed into the climate change movement, addressed the issue of climate change. This followed decades of scientific discovery and data collecting (Christianson, 1999; Davenport, 2008; Soroos, 2002; Zillman, 2009). The scientific community had brought the issue to the attention of policymakers who had explored the issue years before social movements were concerned with climate change (Bolin *et al*, 1986; Dessler and Parson, 2010, 23–4; Falkner *et al*, 2011). It was not until international dialogue around climate change was beginning to form that environmentalists mobilized.

In 1989, environmental organizations in Europe and the United States developed a coordinated network on the issue of climate change known as the Climate Action Network (CAN). The World Wide Fund for Nature (WWF), Environmental Defense Fund, and Greenpeace International led the way and established CAN out of a meeting in 1989 in Hanover, Germany (Weart, 2011, 72). They helped to form regional networks in Europe (Climate Network Europe) and the United States. CAN originally started with 63 ENGOs from 22 countries (Newell, 2000, 126; Rahman and Roncerel, 1994, 246) and, as a loosely coordinated network, served as a means through which environmental organizations could discuss issues of climate change, share strategies on influencing international negotiations, and develop a common platform.

Collectively, these ENGOs' first goal was to influence policymakers who were set to meet at the Second World Climate Conference (WCC-2) in 1990 and ensure that steps were taken toward the creation of international commitments to mitigate climate change.[5] The WCC-2 was composed of two sections. The first section resulted in a statement by scientists regarding the risks posed by climate change. The second consisted of an agreement made by delegates from 137 states and the European Community, where Europe advocated for carbon dioxide (CO_2) to reach 1990 levels by the year 2000 as a minimal basis for agreement. After intense debate, the Ministerial Declaration did not include any specific targets. Instead, broader principles of equity and sustainable development were agreed to, developed countries were urged to establish targets to limit greenhouse gas emissions, and a call was made to establish a framework treaty that included commitments to be adopted by the 1992 United Nations Conference on Environment and Development (UNCED), also known as the Earth Summit.

Prior to Earth Summit, NGOs also worked to inform policymakers and the general public about the issue. In 1990, Greenpeace published a book entitled *Global Warming: The Greenpeace Report*, which described the science of climate change and its possible consequences. In India, the Centre for Science and Environment was working on its publication *Global Warming in an Unequal World* (Agarwal and Narain, 2003 [1991]). At the same time, Friends of the Earth was using its democratic structure to raise public awareness through local groups in countries where it had significant membership (Rahman and Roncerel, 1994, 245).

UNFCCC and the Earth Summit

Although CAN was unable to achieve its goal of calling for specific targets during WCC-2, the statement produced at the conference was strong enough to lead the UN General Assembly to establish the Intergovernmental Negotiating Committee (INC) for a Framework Convention on Climate Change at the end of the year. The INC was composed of five meetings to prepare the framework (eventually to become the UN Framework Convention on Climate Change or UNFCCC), which would be negotiated at the Earth Summit.[6] CAN attempted to influence discussions at these meetings, primarily through the regular publication and circulation of a daily news journal they created: *ECO*. The journal summarized the various debates that occurred during the conference while presenting alternatives and suggestions (Rahman and Roncerel, 1994, 249). *ECO* was described as 'the most widely read source of information on the negotiations' (Dowdeswell and Kinley, 1994, 129; also see Betsill, 2008).

In addition to *ECO*, ENGOs also communicated directly with policymakers, some of whom granted them regular access. The issue of climate change, being so complex and multifaceted, made it easier for delegates seeking expert opinion to welcome input from these organizations (see Rahman and Roncerel, 1994, 251). Such direct participation became formally incorporated into the process. The INC Bureau provided an official platform for NGOs to make a statement to the full plenary at each INC session (INC, 1991, 11) in which a (heavily debated) consensus opinion amongst NGOs was presented (Rahman and Roncerel, 1994, 251–2; for more see Rahman and Roncerel, 1994, 252–5). In addition, the first INC included the circulation of draft texts prepared by NGOs (Bodansky, 1994, 62). Despite this access, ENGOs were never truly at the heart of the negotiations (Faulkner, 1994, 231). To the extent that they did have influence, they benefited from cooperation. This realization led to increased membership in CAN and the establishment of additional regional networks in the global South.[7] However, despite increased willingness to participate by NGOs from the global South, access to policymakers was still decidedly greater for Northern NGOs.[8]

The major goal of ENGOs in the INC negotiations was to ensure a strong text for the convention, inclusive of the concept of ecological limits and binding emissions reduction commitments. These ideas were opposed by major Northern delegations, particularly the United States (Rahman and Roncerel, 1994, 258). Progress did not come easy for ENGOs. Many were frustrated with the lack of will on the part of many developed countries, but experienced negotiators reassured them that 'this was only the natural working of the UN system' and 'only the beginning of a process' (Rahman and Roncerel, 1994, 258; also see Paterson, 1996, 53). Despite some impacts on the negotiation process, the Climate Convention that was drafted included many features some ENGOs vigorously opposed.[9]

The Earth Summit included 172 countries, over 115 heads of state or government, 9,000 members of the press, and over 3,000 NGO representatives (Adams, 2001, 80). The central outlet for NGO participation was the Global Forum, which, although sometimes referred to as the counter-summit, was organized with the help of the UNCED secretariat. While the forum helped to bring NGOs closer together and generate publicity, some felt that it provided a distraction from the international negotiations taking place at the main conference over 20 miles away (Youth Sourcebook on Sustainable Development, 1995).[10]

The Earth Summit became the vehicle for countries to sign up to the UNFCCC, which only included a general commitment to stabilize emissions at 1990 levels by 2000. The general commitment was not specific enough to make the convention legally binding. This was due to the United States' refusal to accept firm emission reduction targets. The United States' economic concerns about cutting emissions rubbed off on Europe, which grew to see that such a convention, without US support, would reduce their economic competitiveness (Little, 1995). Nevertheless, the UNFCCC provided the framework for negotiating stronger commitments in the future.

Conferences of the Parties and the Kyoto Protocol

Following the UNFCCC, a new round of negotiations was launched in order to strengthen the meagre commitments developed at the Earth Summit. The first Conference of the Parties (COP 1) was held on 28 March to 7 April 1995 in Berlin. There, many ENGOs advocated for a draft protocol produced by the Alliance of Small Island States (AOSIS) which called on Annex I countries (OECD countries and some countries with 'economies in transition') to cut CO_2 emissions by 20 percent below 1990 levels by 2005. However, the draft protocol was met with reservation by both developing and developed countries. Instead, 72 developing countries known as the 'Green Group' and NGOs produced another draft during the conference. Nevertheless, a compromise on this draft could not be reached until the last night of the conference when delegates agreed that the previous commitments were inadequate and post-2000 commitments of Annex I countries should be strengthened, but they did not agree to any additional commitments. This became known as the Berlin Mandate, which ENGOs considered to be '"soft" at best' (Earth Negotiations Bulletin, no date a).[11]

At the same time as attempting to influence the negotiations from inside the COPs, NGOs, particularly Japanese NGOs, mobilized outside of the conference as well. A network of NGOs formed known as the Kiko Forum, which grew from 46 to 225 organizations by COP 3. The network organized 750 public workshops in 1997, along with 'human chain' protests, a petition to the prime minister signed by 750,000 people calling for greater leadership on the issue, an 'eco-relay', and a 20,000-strong demonstration during negotiations (Reimann, 2002, 179–80).

During the talks at COP 3 in Kyoto, additional commitments were negotiated. Commitments for developed countries in the Kyoto Protocol grew out of a compromise between the US and the EU in the final

Table 2.1 Kyoto emission reduction targets by country

Country	Target
European Union	–8%
US	–7%
Canada, Hungary, Japan, Poland	–6%
Croatia	–5%
New Zealand, Russian Federation, Ukraine	0
Norway	+1%
Australia	+8%
Iceland	+10%

hours of negotiation. Individual countries would have individual commitments (see Table 2.1). In aggregate, this meant a reduction of 5.2 percent of greenhouse emissions below 1990 levels during the 2008–12 period, a far cry from what CAN had called for when they supported a 20 percent reduction by 2005 as outlined in the AOSIS proposal.

ENGOs failed to influence the framing of the issue and the overall outcome of the agreement, but they were able to influence certain national delegation positions that affected discourse and tinkered with the balance of power during the negotiations (see Corell and Betsill, 2001, 98; Betsill, 2008, 60–1). First, some believed that the public pressure that was building up thanks to environmentalists in the United States forced Vice President Al Gore to attend the Kyoto meeting, which was viewed as a turning point in the negotiations. Second, Gore's speech at COP 3 included an impromptu message to the US delegation to be more flexible on their position in order to reach an agreement after representatives from two ENGOs with close relations to the vice president had lobbied him on the issue and *ECO* berated the United States for their inability to compromise (Betsill, 2008, 53). ENGOs also influenced EU delegations which were more interested in appearing to be environmentally friendly, making it easier for ENGOs to gain traction and pressure EU delegates to take a strong stance, allowing them to win concessions from the US in order to strike a deal.[12]

International climate policy after COP 3

The Kyoto Protocol was open to signature in 16 March 1998, and in one year it had 84 signatories, including the United States, Russia, Canada, Australia, and the European Union. Signatories would then be required to ratify, accept, or approve the treaty within their own national government. The protocol would not come into effect unless 55 countries, which accounted for at least 55 percent of the CO_2 emissions of Annex I countries, had ratified, accepted, or approved the protocol. This took considerably longer to achieve. Prior to COP 3, the US Senate unanimously passed the Byrd-Hagel Resolution, which stated that Congress would not support the ratification of a treaty, support which was required by US law, unless it included commitments for developing countries. As the Kyoto Protocol did not include such commitments, it became nearly impossible for the treaty to be ratified by the United States, which was responsible for about 35 percent of Annex I emissions.

The COPs that followed Kyoto were focused on developing rules and regulations to bring the protocol into operation, ensure its compliance, and hammer out details, and NGOs attempted to influence these discussions.[13] Not only did ENGOs engage as consultants and lobbyists during these post-Kyoto COPs, they also engaged in protests and demonstrations as a way to signal their discontent with the pace of negotiations and climate change mitigation. During COP 6, at The Hague, CAN and the e-activist website Avaaz.org launched the Fossil of the Day Award, which was given to the country who performed the worst during each day of the negotiations (www.fossiloftheday.com). The Dutch Friends of the Earth, known as Milieudefensie, also held 'The Dike' event where 6,000 protesters used 30,000 sandbags to build a 1.5 meter high, 400 meter long dike around the conference venue (European Commission, 2014). In the run-up to COP 6, a new network of organizations that took a more radical stance on the climate change negotiations formed. The network, known as Rising Tide, strongly opposed market-based solutions to climate change mitigation. They protested and published pamphlets during COP 6 and, after negotiations broke down, attended COP 6–2 in Bonn in 2001, where activists held a 300-strong critical mass bike protest, protested carbon credits at a conference meeting, demonstrated by having 500 people lock arms to create a human chain, and dropped a large banner from a crane outside the conference center and later unfurled another banner inside the conference center (SchNEWS, 2001).

Mobilization increased in 2001 when George W. Bush took office as president of the United States, leading the US government to openly reject the Kyoto Protocol in 2001. Greenpeace launched the short-lived 'Ratify Kyoto Now' campaign, and WWF launched a campaign with the same intentions in 2002. However, the lack of progress led to the abandonment of campaigning to ratify Kyoto in the US by some ENGOs (Gulbrandsen and Andresen, 2004, 71). The US rejection meant other major emitters were required to ratify the protocol.

The Kyoto Protocol finally came into effect in February 2005, after it was ratified by Russia a few months prior. By this time, Russia's president, Vladimir Putin, had consolidated power to the point that the final discussion over ratification was seen to be his (Tipton, 2008, 69). Some have viewed his belittling of the environmental movement and the lack of access his government granted to ENGOs as having stalled the ratification of the protocol for two years. Although NGOs were sidelined, they continued to campaign for Russia to ratify the Kyoto Protocol. WWF-Russia itself published over 100 articles in favor of the treaty, while Greenpeace Russia ran a website entitled 'Kyoto, yes!' (Henry and Sundstrom, 2007, 52). Other NGOs provided senior policymakers with information on the issue and attempted to convince them that the Kyoto Protocol would provide benefits to the country. With international attention on Russia, Greenpeace International worked with Greenpeace Russia to deliver Putin 10,000 letters calling on him to ratify the protocol, although public concern in Russia was rather low (Henry and Sundstrom, 2007, 53). ENGOs, particularly the national branches of WWF, also applied pressure on other country leaders such as Germany's Schroeder and France's Chirac to push Putin to sign (Henry and Sundstrom, 2007, 58). At the same time, Putin was able to extract concessions including EU support for joining the World Trade Organization, along with possible financial incentives that were available through the emissions trading scheme due to the reduction of emissions that occurred incidentally following the collapse of the Soviet Union (Tipton, 2008). The ENGO Environmental Defense facilitated these negotiations between Russia and the EU (Henry and Sundstrom, 2007, 59).[14]

Only a few years later, at COP 13, tensions flared as Japan and Canada both expressed concerns with the protocol. At the same time, the US and Australia were devising alternative plans to deal with climate change focused on voluntary targets while ENGOs argued that 'the Kyoto Protocol is the only game in town' (Greenpeace, 2007a). CAN Canada was aware of the importance of the Canadian delegation in negotiating a framework for commitments after 2012 and worked to pressure their government to take action at COP 14 in Poznan, Poland. In order to build this pressure, CAN Canada, along with Greenpeace and others, commissioned a poll asking Canadians their views on climate change, which showed that '[n]early two-thirds of Canadians want to see Canada take action to tackle global warming despite the economic crisis' (Climate Action Network Canada, 2008a). In addition, indigenous organizations from northern Canada and environmental groups formed a coalition calling on the government to play a larger role in the negotiations and provide funding for Northern Canada to mitigate and adapt to climate change (Climate Action Network Canada, 2008b). However, Canada failed to push for stronger commitments during COP 14 negotiations and received negative feedback from Canadian ENGOs (for example, Climate Action Network Canada, 2008c).

Prior to COP 14, an additional network formed made up of Southern NGOs and civil society organizations concerned with indigenous interests and local problems resulting from climate change, called the Accra Caucus on Forests and Climate Change. That same year, a network called Global Call for Climate Action, or TckTckTck, formed and argued for local and indigenous community involvement in the implementation of the Reducing Emissions from Deforestation and Forest Degradation (REDD) program that was established as part of the Bali Action Plan in COP 13 (Climate Action Network, 2008). In addition, the Climate Justice Network (CJN) – which was established the last day of negotiations at COP 13 to address the North/South divide concerning emissions and effects of climate change and included Carbon Trade Watch, Friends of the Earth International, Freedom from Debt Coalition, Oilwatch, The Indigenous Environmental Network, and World Rainforest Movement (Climate Justice Now!, 2007) – really began heavily mobilizing in 2009 following COP 14, expanding their membership from 30 to over 200.

Copenhagen and beyond

An agreement needed to be made at COP 15 in Copenhagen to extend reduction targets to the next commitment period. Activists came to the conference in droves to call for a strong agreement, even if different groups had different opinions about what a strong agreement would look like. Activists gathered both inside and outside the conference. Copenhagen marked the largest number of participants at any COP (between 30,000 to 45,000), with NGO observers accounting for more than two-thirds (Fisher, 2010, 12; also see Climate Action Network, 2009b).

Outside the conference, between 60,000 and 100,000 participants gathered to demonstrate, some of whom were representing organizations that were also official NGO observers inside the conference. The high turnout was not only due to the importance of Copenhagen in terms of climate change but also the increased salience of global capitalism within the negotiations, with a rising awareness of the discourse on flexibility mechanisms and emissions trading schemes. CJN joined a new group, Climate Justice Action, to engage in protest and civil disobedience akin to the 'Battle in Seattle' against the World Trade

Organization conference and call for 'system change, not climate change' (Climate Justice Now!, 2009). However, this call for action led to reduced access to NGOs inside the conference, with Friends of the Earth International, Avaaz, and TckTckTck having their observer statuses revoked (Fisher, 2010, 15). One analysis found that while lobbying and observing resulted in only low levels of influence, the indirect effect of demonstrations on negotiations via the media could have 'sav[ed] the Copenhagen climate negotiations from a breakdown' (Rietig, 2011, 26). Rietig argued that demonstrative protests helped to convince heads of states to attend the conference on short notice and increase public salience on climate change (Rietig, 2011).

Other forms of mobilization also took place. TckTckTck, Greenpeace, Avaaz and others also organized a letter-signing campaign to US president Barack Obama, calling on him to push for strong commitments. Five hundred and seventy-five thousand signatures were gathered (Climate Action Network, 2009a). In addition, 50,000 postcards called on negotiators to make ambitious cuts to emissions without including nuclear power in their flexibility mechanisms (Climate Action Network, 2009c).

COP 15 ended with a document known as the Copenhagen Accord, a non-binding statement endorsing a continuation of the Kyoto Protocol, acknowledging the importance of preventing an increase in global temperatures of two degrees Celsius, and agreeing to establish new emission targets by February 2010. The Copenhagen Accord did not set out specific or overall targets, and the deadline for targets was later seen as a 'soft deadline'. Following Copenhagen, activism around climate change had significantly reduced.

Since Copenhagen, ENGOs have claimed that annual climate change negotiations have failed to deliver plans for significant cuts to greenhouse gas emissions. While the 2011 COP 17 in Durban was hailed as a success by the conference president because an agreement was made to establish a legally binding agreement by 2015, ENGOs argued that it was too little, too late. They felt progress was so slow that in COP 19 in Warsaw, NGOs including WWF, Friends of the Earth, and Greenpeace walked out of the conference, particularly over the failure of Annex I countries in providing funding for poor countries to adapt to climate change.

Conclusion: the importance of national interests and national policies

The problem of climate change is global in nature, and any attempt to mitigate it will require an international effort. This can be achieved through a United Nations sponsored treaty, as is being attempted in these annual COPs, or through a more voluntary arrangement. However, in either case, the international political dialogue around the issue of climate change has been rooted in national interests. Throughout the meetings of the UNFCCC, it has been the national interests of key parties that have accelerated or stymied negotiations. The United States' position on an international climate change treaty, while always hesitant, was considerably different during the Clinton and Obama presidencies than under George W. Bush. Canada's decision to leave the Kyoto Protocol, and the decisions of Japan, Russia, and Canada not to accept commitments beyond the Kyoto period, were calculations of national interests. Although the international economic crisis affected the calculus over climate change mitigation significantly, it did not do so in the same way across the board. Ideologies, publics, and contexts at the national level were crucial components that drove national delegations' responses to the climate change crisis.

The climate change movement began by attempting to influence international climate change policy;[15] disparate organizations across the global North and South came together to lobby for a strong international framework. Others protested outside the walls of the convention buildings and in streets across the world. Few outcomes resulted from these actions, and mitigation efforts have stalled. Rather than attempting to influence international negotiations, the climate change movement can perhaps play a more important role in addressing climate change at the national level through national policies. In doing so, the movement would not only bring about national-level change but also influence climate change policy at the international level in several ways.

First, the success of national policies to mitigate climate change can embolden stronger commitments in international negotiations, while failing to advance national policies can result in position U-turns during subsequent COPs. Canada pulled out of the Kyoto Protocol in no small part due to its inability to fulfill its commitment of reducing emissions 6 percent by 2012, compared to 1990 levels. Instead, emissions increased by over 30 percent. Had domestic policy been successful in curbing emissions, Canada's international position may have been different.

Second, the success of other countries' domestic emissions reduction policies may influence a country's calculus for international negotiations. The failures of major emitting countries to curb emissions domestically have made it easier for countries such as Japan and Russia to remove themselves from further commitments past the 2012 period. Legislation that demonstrates a country's commitment to mitigate climate change reduces the risks of other countries addressing climate change because the fear of costly unilateral action is reduced.

Third, in addition to influencing international negotiations, success at the national level in advancing climate change policies can incentivize others to do the same even outside of international negotiations. For example, UK's Climate Change Act 2008 has become a model for many countries' own domestic climate change policies.

Fourth, in an anarchic world system, national-level climate change policies may be the only suitable means for real-world emissions reductions. Due to the ease of withdrawing from international commitments, which partially explained the watering-down of the Kyoto Protocol, the strength of international treaties is unlikely to directly result in high levels of actual emissions reduction without additional national mitigation policies. In addition, national policies have the potential to reduce the price of renewable energy or increase the price of carbon-intensive energy worldwide, and it can help national policymaking institutions internalize the importance of climate change and simultaneously increase the public opinion and issue salience of the problem.

All in all, national policies to mitigate climate change are an important component in a global response to a global problem. Part of the failure of ENGOs' attempts to influence international negotiations on climate change is due to the inability of these efforts to play a significant role on the cost-benefit calculations of national delegations. National policies play an important role in those calculations, making an examination of campaigns for national policy increasingly important. Therefore, it is useful to analyze the outcomes of social movements to address these policies, not only to understand social movements as a whole, but to better understand the possibilities of influencing climate change mitigation through national climate change policy.

National climate change policy and the UK climate change movement

The climate change movement in the UK had steadily developed over the course of international negotiations. It consisted of a variety of organizations seeking a wide variety of ways to mitigate climate change, from consumer choices (the Global Cool Campaign, Greening Campaign), to supporting public transport (Campaign for Better Transport), to mass vegetarianism (the Veg Climate Alliance). Groups range in size and in demographics. Some groups, such as the UK Youth Climate Coalition, geared their efforts toward young people, while others, such as Campaign against Climate Change, worked to make connections with labor unions. Some groups addressed national policies, and they did so in several ways.

The climate change movement saw some attempt to tackle climate change via national policy but noted that '[t]he government's policy at the time was to have policy documents every five years...full of tiny, tiny little policies...but no comprehensive view' (Worthington, 2011). At the same time, there was uncertainty that a new international commitment would be set following the Kyoto Protocol, which gave the UK a 12.5 percent emission reduction target below 1990 levels to be achieved by 2012. However, the movement saw an opportunity for progress when in 2004 Prime Minister Tony Blair pledged to make a serious effort to push for an international treaty on climate change as president of the G8 and during the UK's presidency of the Council of the European Union. After some discussion about what kind of policy to propose, Friends of the Earth called for the legislation of annual targets that would result in 80 percent emissions cuts by 2050 (Interview with Tony Juniper, 18 September 2014). While the Friends of the Earth campaign focused on a comprehensive policy, others worked to add small policies that would help to reduce emissions, such as the Lighter Later campaign by the group 10:10, calling on the government to adjust daylight saving time to ensure lighter evenings so that less energy would be used.

Organizations also worked to stop new developments that would lead to increased emissions by the UK. Campaigners protested the company E.ON's proposal to replace and renew an older coal-fired power plant. Eventually E.ON, turned away by the government, shelved plans for the plant. A similar opposition campaign was launched against expansion at Heathrow Airport, where the aviation industry attempted to build a third runway that would significantly increase the amount of air traffic and become the country's single largest greenhouse gas emitter.

Another concern of the climate change movement was to ensure that sufficient funding was being given to projects that would reduce greenhouse gas emissions either through energy efficiency or renewable energy sources. One idea drafted by an environmental think tank caught the eye of a Friends of the Earth campaigner who formed a coalition to pressure the government to establish a green investment bank that would use government funding to leverage private capital for green investments.

The following chapters examine Friends of the Earth's campaign for comprehensive emissions targets, the campaign to stop Heathrow expansion, and a coalition's attempt to deliver a green investment bank. Each represents a crucial feature of climate change mitigation: enacting targets, stopping carbon-intensive infrastructure, and financing low-carbon technologies.

3
Case Histories of Three Climate Campaigns

This chapter looks at the histories of the three cases we will analyze in this book: the Big Ask campaign that called on the government to pass a law creating greenhouse gas emission reduction targets; the campaign against a third runway proposed at Heathrow Airport which, if constructed, would increase emissions from aviation; and the campaign for a green investment bank that would increase the financing for climate change mitigation efforts. Below you will find the history of each campaign, including key actors, tactics, and policy developments. These case histories will provide us with the necessary background to answer the following questions: *What* outcomes did the campaigns produce? *When* were conditions ripe for the campaigns to achieve outcomes? and *How* were the campaigns able to achieve these outcomes?

Campaigning for emissions targets: the case of the Climate Change Act

Friends of the Earth's attempt to pressure the government into passing a climate change bill was known as the Big Ask campaign. The proposed bill was the brainchild of FoE. While the Conservative Party and the Labour Party included emissions reduction pledges in their party manifestoes prior to the May 2005 General Election, neither proposed legislation. Instead the government had a scattered approach to climate change mitigation, something FoE hoped to change. FoE's proposed bill was substantially stronger than the political parties' pledges of 60 percent emissions reduction by 2050. Instead, FoE called for 80 percent reduction including emissions for international aviation and shipping, along with annual targets that would make each elected government accountable for mitigation. The Big Ask campaign took a huge effort and a significant portion of the organization's resources. FoE utilized a wide range of tactics to convince policymakers to adopt the legislation, built predominantly around a strategy of mobilizing the public to lobby their representatives.

Following strong statements about climate change by Prime Minister Tony Blair, Friends of the Earth felt it was time to push for a big climate change campaign. After considering several options, they settled on the idea of a national climate change bill and began diverting resources from other campaigns to the new Big Ask campaign (Interview with Tony Juniper, 18 September 2014). FoE's first step for the campaign was to assemble Members of Parliament (MPs) from different political parties to propose a bill in order to show cross-party consensus. FoE had regular communication with key government and opposition policymakers concerning the environment and did not find it difficult to organize a meeting between former Labour environment minister Michael Meacher, former Conservative environment minister John Gummer, and Liberal Democrat environment spokesman Norman Baker (Interview with John Gummer, 26 March 2012; Interview with Norman Baker 12 March 2012). By having this team agree to a draft bill, FoE hoped to apply pressure on the government to adopt its own version of the bill. It was understood that the bill would not become law without the government's support since it held the majority in Parliament. Therefore, the bill's purpose was to increase the salience of climate change in the political and public spheres while offering a ground-breaking but feasible policy option (Interview with Tony Juniper, 18 September 2014). The policymakers sat with FoE and, working with 'unusual enthusiasm' (Interview with John Gummer, 26 March 2012), agreed to a bill that largely resembled FoE's original idea. The draft bill established mandated annual greenhouse gas emissions targets until 2050, including emissions from international aviation and shipping.

On 7 April 2005, Meacher introduced a presentation bill, which gives notice to Parliament that an MP will present a bill on a future date. Shortly after, Parliament was dissolved ahead of the general election in May. The election resulted in another Labour government with Tony Blair remaining prime minister, but with the Labour Party losing a substantial number of seats to both the Conservatives and Liberal Democrats, the two largest opposition parties. On 25 May, about one week after the State Opening of Parliament, FoE officially launched their Big Ask campaign to coincide with the an early day motion (EDM) in support of the FoE-designed bill. An EDM is a tool for MPs to show support for policies and positions proposed by the opposition (the political parties not in control of the government) and the backbench (MPs who do not hold a position in the ministry). The campaign's next efforts were invested in developing cross-party support for the EDM in order to pressure the government.

The campaign's first public appearance came in the form of a media blitz. Friends of the Earth acquired the support of celebrity Thom Yorke, frontman for the alternative rock band Radiohead, who appeared alongside FoE director Tony Juniper for a press conference near Parliament. They then promoted the campaign in TV and radio interviews while FoE disseminated a press release that resulted in 69 newspaper articles, according to the organization (Friends of the Earth, 2005). This big push coincided with and contributed to a significant number of signatories to the EDM from MPs who did not hesitate to show their support.

After this initial push, FoE scaled down its campaigning. However, small and often local events still occurred. Campaigners joined the Camden Green Fair in June to promote the campaign and call on members of the public to act. They collected approximately 1,000 signatures for a Big Ask petition and sent out over 400 postcards addressed to MPs that FoE had made for the campaign which called on MPs to sign the EDM. That same month, local groups held a day of action in which they polled residents on their views about climate change and government action as a means to attract new participants and local media attention. Local chapters also lobbied their MPs and distributed a FoE-produced video advertisement at local cinemas. This last tactic eventually resulted in the short advertisement being seen by approximately 300,000 people in the summer of 2005.

On 13 July 2005, Meacher introduced the Climate Change Bill to the House of Commons. The bill set emissions targets starting from 2010 at a rate of reduction of 3 percent annually. The bill proposed emission targets for specific sectors and called for a government strategy to curb emissions by incorporating the 'polluter pays' principle (Climate Change Bill, 2005). If targets were not met, the bill called for a reassessment of the emissions reduction strategy. If emissions increased by 10 percent beyond the annual target, the bill called for a select committee to consider a recommendation to reduce the prime minister's salary by up to 10 percent. The same penalty would be applied to cabinet ministers if their sectoral targets were surpassed.

At this point, the early day motion (EDM 178, 2005) had been signed by over 200 of the 646 MPs (see Figure 4.1). Like most opposition or backbench bills, it was soon dropped as, despite the large number of signatories, the bill lacked a majority that could stand up to the whips (who ensure party discipline) in both the Labour government and the Conservative opposition. Nevertheless, FoE knew this was only the beginning of the campaign and they felt confident. The Big Ask campaign went into temporary abeyance at this stage, but more actions were later planned.

The campaign reignited in March 2006 when Stop Climate Chaos (a coalition of over 100 NGOs including FoE that came together to tackle climate change) co-sponsored the 'Carbon Speed Dating' event alongside FoE. The event was held near the Palace of Westminster, which houses Parliament, and it served as a forum that brought together constituents and MPs to discuss the Climate Change Bill. The event was attended by 650 activists and 60 MPs including the new Conservative Party leader, David Cameron.

The next campaign events included a series of live performance concerts promoting the campaign, known as Big Ask Live. Bands performed at sold-out concerts,[16] attended by Cameron and Environment Secretary David Miliband, while volunteers publicized the Big Ask campaign and fundraised. FoE was eager to show policymakers that the campaign had the support of a broad spectrum of the public. In addition to organizing Big Ask Live, they commissioned an opinion poll that found that 75 percent of people between the ages of 16 and 64 favored a new law calling for annual carbon emission reductions. MPs showed additional support by sending Prime Minister Blair a letter signed by both the Conservative and Liberal Democrat shadow environment ministers and others calling for the Climate Change Bill to be taken up by the Labour government.

On 1 September, the Conservative Party, hoping to 'green' their image, officially supported the FoE bill with party leader David Cameron appearing alongside Tony Juniper at a press conference and calling on the government to announce the bill in the Queen's Speech, where governments introduce new legislation. Shortly after, FoE initiated a series of events labeled 'the Big Month', which centered on organizing constituents to contact their local MPs and lobby them on the Climate Change Bill more directly than by sending letters or signing petitions. The strategy moved from getting the public to pressure MPs to having MPs pressure the government. This involved campaigners meeting face to face with their representatives and asking them their opinion on the subject, lobbying them on the bill, and asking them to write a letter to the government calling on them to act.[17] Campaigners communicated with 620 out of 646 MPs during the Big Month, leading to some positive responses (Friends of the Earth, 2006c, 2008a).

FoE was feeling confident, and at the end of the Big Month, environment secretary David Miliband hinted that climate change legislation would be included in the Queen's Speech. Tony Blair's statements on

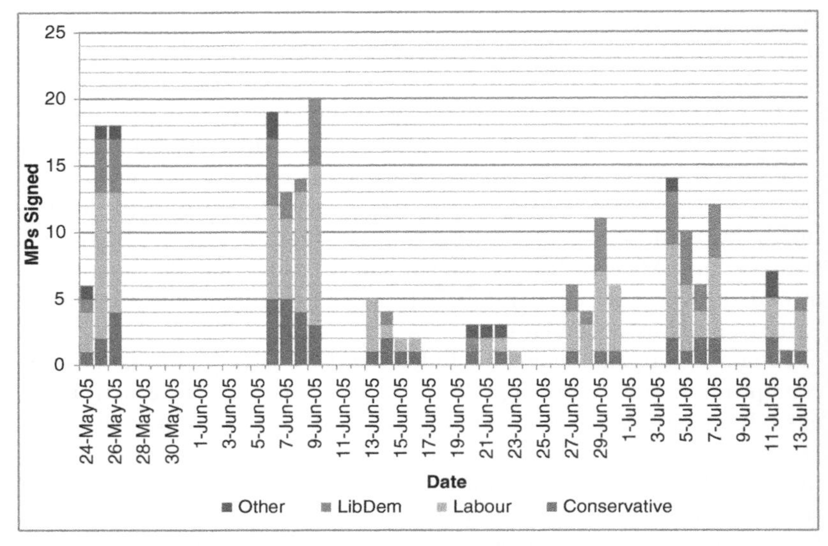

Figure 3.1 EDM 178, signatures by party (24 May–13 July 2005)

Note: Gap from 27 May to 5 June is Parliament in recess for Whitsun.

Source: Parliament UK, no date.

the matter were less revealing, stating that 'we will have to wait for the Queen's Speech and the outline of the bills it contains' (House of Commons, 2006, c299). At this point, the EDM had nearly 400 signatories. The Queen's Speech was one month away when MPs increased pressure on the government by writing to the *Times* newspaper to call for the legislation and Cameron published a model climate change bill.

On 4 November, Stop Climate Chaos hosted the 'I count' rally and concert as part of the I-Count campaign 'designed to inspire personal and political action and counter the view that climate change is too big a problem to fix' (Friends of the Earth, 2006d). The demonstration, which included a march starting from the US embassy in London and ending at Trafalgar Square, called on the UK government to take a leading role in tackling climate change including the adoption of a climate change bill. It was the country's largest climate change demonstration up to that point, with between 20,000 and 25,000 participants.[18]

On 15 November, a climate change bill was announced in the Queen's Speech. The campaign had achieved one of its first goals but waited to see the draft bill that the government would produce. The government called on Friends of the Earth climate change campaigner Bryony Worthington to help draft the bill. This draft bill appeared in March

2007, but FoE found it dissatisfying, arguing, '[it] still has some pretty big holes in it, and really fundamentally the biggest problem with it is it's just not ambitious enough to actually make the cuts in carbon dioxide that we need...to prevent dangerous climate change' (FoE senior parliamentary campaigner Martyn Williams, interviewed in Climate Radio, 2007). The campaign relaunched, targeting what FoE felt were the three major inadequacies of the draft bill:

1 The bill called for a 60 percent cut to CO_2 emissions by 2050, whereas FoE argued for at least an 80 percent cut by that time. The 60 percent target was already backed by the government following a recommendation by the Royal Commission on Environmental Pollution, although it was not legislated in Parliament. FoE claimed that new scientific data required this target to be reconsidered.
2 Instead of annual targets, the government's draft bill called for five-year carbon budgets. FoE was concerned that five-year budgets would not lead to the levels of accountability needed for continuous efforts in reducing greenhouse gas emissions. They instead advocated for annual targets that would ensure each elected government would do its job to reduce emissions each year.
3 International aviation and shipping were excluded from the emissions targets.

It did not take long before the issue of annual targets was resolved. During a discussion with the Joint Committee on Draft Climate Change Bill, senior parliamentary campaigner for FoE Martyn Williams stated, 'We are perfectly happy to admit we have changed our position', realizing that a five-year budget was more practical as natural variability in emissions could not be controlled by policymakers (in Joint Committee on the Draft Climate Change Bill, 2007a, ev 58). FoE continued advocating for strong annual reporting. During his testimony, Williams suggested that the five-year budgets could simply be viewed in annual terms by dividing the target by five, 'but it does not necessarily have to be that way' (Joint Committee on the Draft Climate Change Bill, 2007a, ev 58). The Joint Committee acknowledged this point and recommended an annual indicative target range, stating:

> external organisations will analyse future budgetary targets and calculate the proposed trajectory in terms of annual emissions. These [five-yearly budgets] would become de facto annual targets. Rather than leave this process to external forces, it would be preferable for the Government to agree to indicative annual milestones...against which performance could be assessed (Joint Committee on the Draft Climate Change Bill, 2007b, 26).

The government later amended the bill to include these indicative target ranges.

Regarding the two other amendments, the campaign returned to its old tactic of asking their members and the public to write to the government calling for a tougher climate bill during the three-month consultation period. FoE again relied on celebrity endorsements to spread the message. They recruited director Kevin Macdonald to create the Big Ask Online March and got celebrities such as actress Gillian Anderson and actor/comedian Stephen Fry to advertise the march, asking others to submit videos of themselves calling for amendments to the draft bill. The march received over 1,000 video submissions.

In September, FoE called for a new series of events entitled 'Big Autumn Push' which called on local chapters to lobby MPs, hold public meetings, and increase visibility in their communities in an attempt to pressure the government on the key amendments to the draft bill following the government's consultation. As part of the draft bill, the government called for the establishment of a Committee on Climate Change (CCC) that would advise the government on pertinent issues. Around this time, the government called for a shadow CCC to be established in order to provide information and recommendations in time for the legislation to be implemented.

By this point, Tony Blair had resigned his post as prime minister and Chancellor Gordon Brown became the new Labour Party leader and prime minister. Brown (2007) called on the CCC 'to report on whether the 60 percent reduction in emissions by 2050, which is already bigger than most other countries, should be even stronger still'. The Big Autumn Push included large numbers of training events, public meetings, documentary screenings, streets stalls, and meetings with MPs (see Table 3.1).

In October, Early Day Motion 2233 was introduced by Nigel Griffiths MP, calling for amendments to the government's climate change bill. FoE

Table 3.1 Friends of the Earth activity during Big Autumn Push (Sept–Nov 2007)

	Training Events	**Public Meetings**	**Screenings of *An Inconvenient Truth***	**Meetings with MPs**	**Street Stalls**	**Media Stunts**
# of occurrences	25	46	58	163	445	94

Source: Friends of the Earth Trust Limited, 2008.

supported the EDM, and some local groups lobbied their MPs to sign it. It was on 14 November 2007 that the Climate Change Bill was introduced to Parliament. It did not contain any of the suggested amendments FoE had proposed. However, environmental minister Hilary Benn did state that the government would consult the CCC to advise them on the issues. EDM 2233 had received approximately 100 signatures at this time.

Concerning the issue of international aviation and shipping, FoE sent press releases arguing that the government was lying about their overall emissions reductions because aviation was not included, pointing out that emissions from these sectors accounted for 6.29 to 7.6 percent of the country's total emissions. They did not receive the support they had expected from Conservatives on the issue, but they did receive some from the Liberal Democrats who proposed amendments in the House of Lords. However, the amendment failed easily, with the government stating that international legislation should precede the inclusion of emissions from these sectors and citing concerns that unilateral steps could damage international dialogue. Instead, the government once again turned to the CCC to advise on their inclusion.

When the bill entered the House of Commons, Thom Yorke again acted on the issue, guest-editing a climate change edition of the *Observer Magazine*. Then, under the banner of I-Count and the Big Ask campaign, Stop Climate Chaos and FoE organized a public debate with the new environment minister, Hilary Benn, and shadow ministers Peter Ainsworth and Steve Webb. During the debate, in front of over 1,000 attendees, Benn reiterated that he felt the target needed strengthening but would defer until the CCC's report. Continuing to pursue celebrity endorsements, FoE worked with the band Razorlight, which played a concert promoting the Big Ask campaign.

All of FoE's amendments in the House of Commons had failed, but it was not long before the shadow CCC sent its interim report to Ed Miliband, who had recently become the minister of a new department, the Department of Energy and Climate Change (DECC), which was created during a cabinet reshuffle. The interim report supported an increased target of 80 percent and argued that the bill should either include international aviation and shipping or redistribute those emissions reductions to other sectors (Lord Turner of Ecchinswell, 2008). Miliband then announced that the government would increase the emissions target to 80 percent but also stated that emissions from international aviation and shipping would not be included in the bill but that they would be dealt with at a later time. FoE issued a press release urging the government to include these sectors' emissions. It was reported that, soon after, Miliband 'bow[ed] to pressure from environmentalists and rebel Labour MPs by announcing he will accept an amendment to include these emission sources in the climate change bill' (Hencke, 2008), but the amendment that was approved stated that either these emissions should be included by the end of 2012 or an explanation be provided to Parliament.

On 26 November 2008, the Climate Change Bill received royal assent. On that day, Friends of the Earth stated:

> Friends of the Earth led the campaign for a climate change law, which will oblige the UK to cut its greenhouse gas emissions by 80 per cent by 2050. The new legislation is the first of its kind anywhere in the world – and should put Britain at the forefront of international efforts to tackle climate change (Friends of the Earth, 2008b).

On 19 December 2012, the Conservative – Liberal Democrat coalition government that formed following the 2010 elections agreed to defer the decision for the inclusion of international aviation and shipping yet again. Instead, they argued that the carbon budgets up to 2027 'have already been set to leave headroom for international aviation and shipping emissions, putting us on a trajectory which could be consistent with a 2050 target that includes a share of international aviation and shipping emissions' (DECC, 2012). Despite this minor setback, FoE has framed the Big Ask campaign as a major victory.

Campaigns against carbon-intensive infrastructure: the case of Heathrow's third runway

Prior to the Big Ask campaign, the increasing importance of climate change in the public and political eye did little to rein in the politics of growth. This was particularly true within the aviation industry in the UK. In the 2000s, two major players in the industry joined forces to push for aviation expansion across the UK and particularly at Heathrow Airport, the UK's hub airport, where they called for a third runway to be built. Local opponents to the expansion plans began to mobilize, later jumping scale to become a national campaign. The campaign lasted into 2010 when, following the election of a coalition government composed of the Conservative and Liberal Democrat parties, the third runway was stopped by the government, and the industry withdrew its application for a third runway.

One of the two industry forces was British Airport Authority or BAA (now known as Heathrow Airport Holdings), the operator of Heathrow

Airport as well as other smaller airports. BAA grew out of the Airports Act of 1986 following a wave of privatization during Margaret Thatcher's tenure as prime minister. Starting in 2001, when the government began drafting plans to reduce air traffic congestion in South East England, BAA, together with airliner British Airways who 'would murder to get another runway' (senior aviation industry source quoted in Walters, 2002), lobbied government and publicly called for a third runway at Heathrow Airport. Had the third runway been built, Heathrow Airport would have become the single largest greenhouse gas emitter in the UK.

Despite the threat of increased emissions, at this time, concerns about climate change were not addressed by the local forces that began opposing the runway plans. Instead, local councilors, residents, and community organizations opposed the expansion plans due to threats to local housing and community infrastructure that would likely be demolished, and due to increased noise and air pollution. At the heart of the opposition was a group named HACAN ClearSkies. The group formed in 1999 out of a merger between two community organizations that campaigned against noise pollution from the airport: Heathrow Association for the Control of Aircraft Noise and ClearSkies.

Local campaigners knew that opposing the industry's expansion plan was an uphill battle. The lobbying power of the industry was enormous when it came to aviation policy. John McDonnell, Labour MP for Hayes and Harlington, the constituency that contains Heathrow Airport, stated: 'The aviation industry, in particular BAA, has written both [Conservative and Labour] parties' policies for years, to the extent that BAA's staff are given passes to occupy offices within the Department for Transport offices themselves' (Interview, 21 June 2012). The close relationship allowed the aviation industry to feel confident that expansion would occur where they desired (Interview with John McDonnell, 21 June 2012).

It was not long before ministers began considering a third runway at Heathrow as a way to deal with traffic congestion following the industry's suggestion. Although some argued that a third runway at Heathrow Airport would be technically simpler than expansion at other airports in South East England, it was believed that opposition to another runway at Heathrow would be strong. The local community near Heathrow had recently opposed plans for a fifth terminal, leading to the longest public inquiry in UK planning history, lasting three years and ending in March 1999 (Doherty, 2008, 47). Terminal 5 itself was seen by campaigners as a 'Trojan horse' for a third runway (Friends of the Earth, 2009) despite the government's promise during the Terminal 5 Inquiry that a third runway would not be built. In approving the new terminal, the government also stated that an additional runway would have 'unacceptable environmental consequences' (Friends of the Earth, 2009, 3). When it was announced that a third runway was under consideration, 'some campaigners were shocked, whilst for others it was just another of a long string of broken promises by BAA' (Saunders, 2004, 149).

Although the environmental problems were said to be unacceptable, it did not take long for Roy Vandermeer, the Heathrow Terminal 5 Inquiry inspector, to publicly state that 'environmental impact might alter' (in Harper and Hetherington, 2002), signaling a third runway was not off the table. As more official support was given to the new runway, HACAN ClearSkies (or simply HACAN) began to challenge government figures on projected housing demolitions (HACAN ClearSkies, 2002b) and attacked the government for being 'dreadfully devious' (HACAN ClearSkies, 2002a).

The industry was pushing the issue. Alongside 'quietly lobbying' government (Harper, 2001), British Airways and other airlines were considering contributing to a package to compensate residents for the houses that would have to be bought and demolished and for others that would suffer from increased noise pollution. Depending on the projected figures for air pollution, the number of houses that were forecasted to be destroyed varied from 100 to 400, and later from 12,000 to 15,000 (Walters, 2002; Brown, 2002). News of a potential compensation package did not quell local opposition. Prior to a Department for Transport (DfT) two-day exhibition on the impact of the third runway held near the airport, residents began displaying slogans and posting orange stickers around the community reading 'No Third Runway'. It was not long before collective action was taken. In October 2002, around 600 campaigners, including many elderly locals, demonstrated outside Parliament, protesting the expansion plans. That same month, 300 gathered in Turnham Green, not far from Heathrow, to hold a rally against the runway. Here, local Labour Party MP Ann Keen publicly opposed the third runway and urged her constituents to send her letters in order to strengthen her position (chiswickw4.com, 2002).

Both sides continued to campaign on the issue. BAA argued that the third runway was crucial for economic growth (for example, Marston, 2003), and campaigners continued to protest. In June 2003, protesters marched along the route of the proposed runway and the following month protested outside of British Airways' annual general meeting. The protest included members of a new local residents group, No Third Runway Action Group (NoTRAG). The industry and local campaigners each tried to influence the Labour government as it worked on the 2003

Aviation White Paper that was published at the end of the year. The white paper, a government document that describes future policy on a given issue, green-lit expansion in over 20 airports in the UK, including a third runway at Heathrow, 'provided that stringent environmental limits can be met' (DfT, 2003, 111). The white paper increased the salience of the issue as it sent a clear signal that a third runway was likely to be approved by the government.

Friends of the Earth joined the campaign, and local authorities began seeking legal ways to prevent the expansion. Issues began arising regarding European Community regulations on air pollution, which campaigners hoped would turn the third runway into an 'impossible dream' (Stewart, in Clark, 2004). Nevertheless campaigners continued stirring up local opposition with HACAN and NoTRAG running the Road Show, which consisted of traveling to areas affected by expansion to inform local residents, protesting near Parliament, and holding a community procession against the runway with the help of local Friends of the Earth chapters and Green Party members.

In 2004, the first-ever judicial proceeding against a government white paper was heard after HACAN joined anti-airport expansion organizations Luton and District Association for the Control of Aircraft Noise and Stop Stansted Expansion, along with the London boroughs of Wandsworth and Hillingdon, to mount a legal challenge. The courts ruled in February the following year with mixed success for the campaign. Regarding Heathrow, the judge ruled that a full consultation was necessary prior to the introduction of mixed-mode operations, which is a mixture of take-offs and landings from a single runway that would have the impact of maintaining noise pollution for the entire day. Mixed-mode operation was seen as a way to increase air traffic prior to the approval of a third runway. Although the decision did not directly challenge the third runway, at least one solicitor viewed it as part of a 'huge fight' against the runway (Richard Buxton as quoted in HACAN ClearSkies, 2005a).

Despite hopes that the court challenge would 'mean the need to look again at the whole question of runway provision in the South East' (HACAN ClearSkies, 2005a), the pro-expansion Labour government was not put off by the result. Instead, Chancellor of the Exchequer Gordon Brown soon announced that the government's new transport advisor would be British Airways' chief executive Rodd Eddington, which HACAN felt 'cements links between the aviation industry and government' (HACAN ClearSkies, 2005b). Campaigners retaliated. At 5 a.m. they blasted aircraft noise from speakers right outside Eddington's home in order to give him 'a taste of his own medicine' (Stewart, in HACAN ClearSkies, 2005c).

The battle between the industry and the campaigners was just beginning. A few months later, industry representatives, unions, business associations, and policymakers established the Future Heathrow coalition to promote the expansion of the airport. Campaigners showed up during the coalition's launch event. One campaigner, disguised as a journalist, threw cake on the face of Secretary of State Alistair Darling who was at the event. John Stewart, the head of HACAN, stated: 'It is quite inappropriate that any Secretary of State should so publicly align himself with any pressure group' (quoted in SchNEWS, 2005). This marked the first of a series of industry and government events on aviation expansion that campaigners protested and sought to disrupt. Following this event, members of the direct action environmentalist group Earth First! joined local campaigners in disrupting a three-day aviation conference in London.

> [P]rotesters stormed their way through security into the conference, armed with rape alarms attached to helium balloons which they released into the high ceiling conference suite, timed to coincide with a key note [sic] speech by a senior executive from British Airways. Ironically, each of the six alarms causes 130 decibels of noise – the equivalent of a jumbo jet taking off (Plane Stupid, 2005).

Others later attempted to block the street where delegates' coaches were expected to pass. These events initiated a tide of direct action, with environmental activists promising to use any means to stop the third runway. Even some MPs threatened to use direct action (see Jowit and Hinsliff, 2006).

Much of the campaign's direct action was instigated by a social movement group called Plane Stupid. They formed in 2005 around the issue of aviation's contribution to climate change and regularly held direct action protests. In April 2006, Plane Stupid activists chained themselves to the doors of BAA headquarters, stating 'we opened up a new front in the fight against climate change' (spokesperson Petra Urwin as quoted in Clasper, 2006). This is when the climate change movement stepped in and the third runway really became a climate change issue.

BAA had grown increasingly confident in meeting EU nitrogen oxide and noise pollution standards despite the delay this would cause. However, the industry sensed that the climate change issue would have to be confronted. Days after Plane Stupid's protest, BAA's chief executive Mike

Clasper wrote to *The Guardian* newspaper in order to 'dispel a couple of myths' about aviation's impact on climate change, arguing that aviation was only responsible for approximately 5 percent of greenhouse gas emissions and stating that while there were 'understandable local reasons, [campaigners] should not hide behind bigger arguments about climate change' (Clasper, 2006).

Not everyone was convinced. In fact, in October the All-Party Parliamentary Sustainable Aviation Group launched a report that showed the impact of aviation on climate change, arguing that if policies did not change, the UK would not fulfill the climate change commitments it made when accepting the Royal Commission on Environmental Pollution's recommendation of a 60 percent carbon emissions reduction by 2050 (Cairns and Newson, 2006). The report noted that even based on conservative estimates, carbon dioxide emissions from aviation would quadruple between 1990 and 2050 if policies stayed the course. HACAN praised the report as further reason to oppose the third runway. Campaigners called on the government to act, stating that if it was 'serious about tackling climate change it must abandon its airport expansion plans' (Friends of the Earth aviation campaigners Tony Bosworth as quoted in Milmo, 2006a). The issue of climate change and the campaign itself began to concern the industry. Prior to the Aviation White Paper Progress Report, BAA's new chief executive Stephen Nelson stated: 'I know that the Secretary of State is under pressure from NGOs and environmental groups to turn this progress report into a rethink [but] there can be no U-turn on the air transport white paper' (quoted in Milmo, 2006b).

Indeed, the progress report did nothing to alleviate campaigners' concerns. It did, however, attempt to appease the ENGOs by including an 'emissions cost assessment' that would examine whether the aviation sector covers its 'climate change costs' in order to 'inform decisions on major increases in aviation capacity' (AEF, 2008). At the same time, Chancellor Brown also doubled the air passenger duty, a tax that was hoped would raise the cost for traveling via airplane. Brown's appeasement strategy did not work, as it was opposed by airlines and travel groups, which viewed it as 'a complete U-turn of government policy', while environmentalists stated that it did not go far enough to be effective and would not deter air travel (Done and Blitz, 2006).

The Labour government's middle-ground approach failed to placate protesters, who continued with a series of actions in 2007. At this point, 'the issue rose to prominence and it became the real...iconic front line for the climate movement' (Interview with Joss Garman, Plane Stupid,

4 July 2012). Large environmental organizations became increasingly involved as climate change became the focus of opposition. For example, Greenpeace constructed 'impromptu ticket exchange booths' where they exchanged domestic British Airways tickets for train tickets, which they explained were more climate-friendly (see Greenpeace, 2007b). Greenpeace later helped to interrupt the transportation minister's airline industry conference. Plane Stupid also continued to protest, stopping traffic outside of the Department for Transport and later chaining themselves to the offices of BAA headquarters to mark the 60th anniversary of Heathrow Airport.

It was in August when media attention surrounding Heathrow and the campaign exploded, stretching across the globe. A loosely networked group of activists known as Camp for Climate Action, or Climate Camp, set up a temporary **protest camp** near the airport. Even before their arrival, they attracted media attention as the threat of a protest camp led BAA to seek an injunction against the protesters in court. The camp, being open to anyone and having fluid day-to-day membership, made it difficult to name individuals in the injunction. Therefore, the initial request was to include all members of campaigning organizations and their affiliates. This included the members of organizations affiliated with AirportWatch, a network of local anti-airport expansion organizations and national environmental organizations that was created by HACAN's John Stewart. A detail that appeared to have been overlooked by BAA was that AirportWatch included organizations such as Royal Society for the Protection of Birds, Campaign to Protect Rural England, Friends of the Earth, and the National Trust. Their combined membership was over six million people. In fact, the judge herself stated that she was a member of at least three of the organizations (Murray, 2007, 24). While the injunction was granted, it was scaled down significantly to include only Plane Stupid members, John Stewart, and NoTRAG chair Geraldine Nicholson (Stewart, 2010). The extreme nature of the original injunction request resulted in national media coverage for the campaign, but the ruling meant Climate Camp had to move the camp just outside of the perimeter of the injunction to allow Plane Stupid members to participate.

As soon as the camp came to Heathrow, media attention went global. This was the effect of the unpredictable nature of the protest camp and the possibility that activists could attempt to shut down a major hub airport, although protesters repeatedly stated that they had no plans to disrupt passengers (Interview with Hannah Garcia, 16 July 2012). The camp was used as a hub for activists who gathered to hold meetings and press conferences, provide workshops and trainings, and plan demonstrations that occurred in the last two days of the camp. These actions included protests, road blockades, occupations, banner drops, and actions in solidarity with striking cargo workers and occupied Palestine. While the camp was seen as an illegitimate and unacceptable form of protest by many policymakers (Interview with Hannah Garcia, 16 July 2012), some MPs who were opposed to the third runway showed their support by coming to the camp, including Liberal Democrats Vince Cable and Susan Kramer, as well as local MP John McDonnell, who had spent years opposing airport expansion in a variety of ways (Interview with John McDonnell, 21 June 2012; Stewart, 2010, 31).

While the Liberal Democrats maintained their opposition to the third runway, the Conservatives, under their new leader, David Cameron, were not yet convinced. Cameron's Quality of Life Report came out shortly after the protest camp and criticized airport expansion, but Conservative Party members did not allow this to become party policy. Some Conservatives felt that a U-turn on airport expansion would jeopardize the party's traditional economic approach. At the same time, some campaigners felt that the industry was taking the Conservative Party's support for granted (Stewart, 2010, 29).

While the Conservatives were still divided on the issue, the Labour government went ahead with a three-month public consultation on a third runway and an additional terminal the third runway would require. Campaigners used the consultation as an opportunity to mobilize. NoTRAG and HACAN held over 40 public meetings and organized oppositional exhibitions at the same hotels and on the same day as an official DfT exhibition, which 'shocked and dismayed the Department for Transport and the aviation industry' (Interview with John Stewart, 9 December 2011). Shortly afterward, Greenpeace discovered details of collusion between BAA and the government regarding the consultation, including that BAA had written part of the consultation document, provided data on pollution and noise that was used by the government, and set up a joint body with the government called the Heathrow Delivery Group (Greenpeace, 2007c). At around the same time, a new coalition formed called Stop Heathrow Expansion, representing large established groups such as the National Trust and the mayor of London's office (Interview with Joss Garman, 4 July 2012). The increased mobilization and evidence of collusion, however, did not diminish the government's resolve.

At this point, the campaigners had few options. While the government admitted that it would scrap the third runway proposal if the consultation sided with the opposition campaign and that campaigners could challenge the evidence if the consultation supported a third runway, neither of those seemed likely. The campaigners felt that the consultation was going to support a third runway due to historic links between the industry and government and the Labour government's steadfast position thus far. If the consultation supported a third runway, the campaigners did not have the resources to conduct the necessary research to challenge the consultation.

Campaigners were not discouraged. Instead, this added pressure spurred additional action. Plane Stupid activists protested by interrupting a Transport Select Committee meeting to scrutinize DfT policy on BAA and handed out a report on aviation emissions and climate change produced by the Tyndall Centre for Climate Change Research. NoTRAG continued to inform local residents on expansion details and attended the annual climate change rally that was organized by the Campaign against Climate Change. Local authorities joined campaigners and formed 2M, representing two million residents near the airport who were opposed to the expansion plans. HACAN and NoTRAG organized a rally that persuaded other council leaders to join (Stewart, 2010, 18), eventually expanding 2M to reach over five million residents and over 20 local authorities. Friends of the Earth called on Virgin Atlantic airlines owner Richard Branson to oppose the third runway while Greenpeace climbed atop a British Airways aircraft, dropping a banner that warned of a 'Climate Emergency' and called for 'No 3rd Runway'. Plane Stupid protesters went even further, climbing the roof of the Houses of Parliament and displaying banners that read 'BAA HQ', with the HQ for headquarters. An advertisement paid for in part by Greenpeace appeared in the *London Evening Standard* stating that the major mayoral candidates were all opposed to the runway. Additionally, HACAN and NoTRAG organized a rally at Central Hall in London just before the consultation finished. The venue was filled to its 2,500-person capacity with an overflow of approximately 500. Rally attendees included the Liberal Democrat's new leader, Nick Clegg, and Labour MPs Ann and Alan Keen.

It would take the government some months to respond to the consultation documents, having received a total of 69,377 responses, with nearly 2,000 more responses arriving too late for consideration. Of the responses, 62 percent were categorized as campaign postcards and petitions. While HACAN put little effort into promoting the consultation as a vehicle for stopping the third runway, other campaigners and local authorities did reach out to the public to submit consultation responses, often in the form of postcards or pro-forma letters and emails. Of the campaign responses, Hillingdon Council (14,994), Greenpeace (6,698), Wandsworth Council

(4,569), Hammersmith and Fulham (4,270), and NoTRAG (4,051) represented the campaigns with the highest levels of submissions. The countermobilization by British Airways (1,494) and Future Heathrow (3,769), also included in the DfT's coding for campaign responses, produced fewer but still significant numbers of responses (DfT, 2008).

While the government was busy assessing the consultation documents, HACAN and NoTRAG released a report they had commissioned which questioned whether or not Heathrow was vital to the British economy, arguing that previous economic studies used best-case scenarios and contained methodological flaws (see Boon *et al*, 2008). The campaigners went on to use these findings to lobby policymakers to show them that a third runway was not an economic necessity. At this time, campaigners were given indications from local authorities that the Conservative Party was interested in turning over a new green leaf and becoming a more environmentally friendly party (Interview with John Stewart, 9 December 2011). John Stewart saw the report as necessary to bring the Conservative Party on their side of the issue considering the importance the party placed on economic growth (Interview with John Stewart, 9 December 2011).

Protests continued. A large **flash mob** was held in March 2008, at the grand opening of Heathrow Terminal 5, with hundreds of people participating, wearing identical 'Stop Airport Expansion' T-shirts. Then, at the end of May, 3,000 protesters joined the festive Make-a-Noise Carnival, forming a giant 'NO' on the field in opposition to the runway. Around this time, support for the runway among policymakers began to slip. The Environment Agency and the independent Sustainable Development Commission came out in opposition to the runway; a Greenpeace survey found that 18 Labour MPs in London were against expansion; a survey by Alan and Ann Keen found that 90 percent of the London borough of Hounslow opposed the runway. It did not help the aviation industry's cause that Conservative MP Justine Greening, long opposed to the runway, had obtained documents through a Freedom of Information Act request which became part of a *Sunday Times* report showing further collusion between the Department for Transport and BAA (Personal Correspondence with Justine Greening, 17 September 2014).

Then, in June 2008, the Conservative Party under Cameron's leadership decided to make a policy U-turn and opposed the third runway, promising to block proposals for its construction if they formed a government following the 2010 elections. The policy reversal shocked many. British Airways chief executive Willie Walsh called Cameron's decision 'extremely insulting' and accused him of backing 'flawed arguments' (as quoted in Milmo, 2008a). Cameron used the economic report commissioned by HACAN and NoTRAG to support the party's decision, along with a report by business group London First, which argued that Heathrow needed to focus on customer service rather than expand to be financially successful. The Conservative Party later called for the expansion of rail services connecting London to Manchester and Leeds, to which British Airways stated that the Conservative Party was 'becoming an unlikely darling of the green lobby' (Milmo, 2008b).

This turn of events threatened the Labour Party. In addition to being out-greened by the Conservatives, Labour was set to lose between ten and 20 seats in the southeast of England if they continued to support expansion, according to Labour MP McDonnell (Interview, 21 June 2012). Also, former Labour environment minister Michael Meacher (2008) expressed concerns about greenhouse gas emissions targets being exceeded if the third runway was approved. Labour rebels began to make noise. This included environment secretary Hilary Benn, energy and climate secretary Ed Miliband, deputy leader of the Labour Party Harriet Harman, and David Miliband, who had become foreign secretary in 2007 when Gordon Brown took Tony Blair's spot as prime minister after the latter stepped down. These ministers argued with Gordon Brown on the basis that Heathrow expansion was costing them their green credentials and marginal seats near the airport and was putting pressure on their flagship climate policy, the Climate Change Act 2008.

A decision on the runway that was scheduled to be made in December was delayed. It was speculated that Ed Miliband had convinced Prime Minister Brown to reconsider the runway for environmental reasons. The one-month delay gave campaigners optimism and frightened the aviation industry. Willie Walsh (2008) wrote to *The Guardian* urging the climate campaigners to move their concerns away from Heathrow, arguing that stopping the third runway 'will not reduce absolute emissions one iota'.[19] Labour rebels did not back down. McDonnell argued that concerns over the runway would result in politicians 'standing [as candidates] against [Labour] either under the same party banner but refusing to support the manifesto position on it, or serving independently' (Interview, 21 June 2012).

It was acknowledged that regardless of the Labour government's position, it would be the next government that determined the fate of Heathrow expansion. A judicial review would delay the application process until after the elections. With the Conservatives and Liberal Democrats set to block the expansion if they formed a government, which appeared likely based on polls, campaigners were hopeful that the runway would be stopped. Nevertheless, they continued to mobilize and demonstrate, although often in a more festive way. Climate Rush (a group of direct action campaigners inspired by the suffragettes) and the local branch of the grassroots Women's Environment and Climate Action Network hosted a 'sit-in dinner at domestic departures' in Heathrow Airport's Terminal 1 where 700 participants played musical instruments, wore costumes, displayed banners, and ate sandwiches and cupcakes.

Although things were looking hopeful for the campaigners, they continued to plan for the worst. If the runway was approved, they would need another strategy to stop it at the construction phase. The need for such a strategy led to a Greenpeace action known as Airplot that saw comedian Alistair McGowan, actress Emma Thompson, Greenpeace director John Sauven, and Conservative and Liberal Democrat MPs come together to purchase a one-acre plot of land where the third runway would be constructed in order to divide up and sell off small parcels to various people to slow the construction process. The tactic would force the government to issue compulsory purchase orders and seize the land from a large number of owners. It was not long before 5,000 owners were found.

Amid the protests and media attention, some cabinet members attempted to delay the decision on the runway that was expected on 15 January 2010, calling for a more detailed environmental report, particularly concerning climate change commitments. However, transport secretary Geoff Hoon approved the third runway at Heathrow, which allowed BAA to proceed to develop detailed plans for construction. Some left-leaning commentators saw this as Labour's 'final betrayal', adding, '[i]t's almost enough to make you vote Conservative' (Monbiot, 2009). John McDonnell was so outraged at the decision that he seized the mace in the House of Commons, representing the royal authority of Parliament, placing him in contempt of Parliament.

The decision approved a third runway, but the opposition led by Hilary Benn and Ed Miliband, which came to be known as the 'Milibenn' tendency, did win concessions. These concessions included increases to public transport investments, the allocation of new aircraft slots to lower-pollution-emitting planes, a new greenhouse gas emissions target for aircraft, and an initial cap on third runway flights to half capacity. While these concessions amounted to 'half a runway' (cabinet member, quoted in Kirkup, 2009) campaigners were not satisfied. New calls were made for direct action to take place. A protest was held by CaCC near the prime minister's office on Downing Street in London and another flash mob was held at Terminal 5. Climate Rush later chained themselves to the gates of the Palace of Westminster in protest during the Conservatives' opposition day in the House of Commons. On that day, the Conservatives held a non-binding parliamentary vote on Heathrow. Despite it being non-binding, the Labour government issued a three-line whip, forcing Labour MPs to attend and vote with the government or else face punishment. Nevertheless, some Labour MPs resigned their junior ministerial posts in order to vote against the third runway, contributing to 'the biggest rebellion on an opposition motion since Labour came to power in 1997' (HACAN ClearSkies, 2009). Still, the rebellion was not big enough for a majority, as the government was joined by some opposition Conservatives to defeat the motion.

Protests continued up until the elections. Protests occurred outside Downing Street, during speeches by Ed Miliband and Geoff Hoon, and during a large music festival in Glastonbury. Greenpeace invited architects to design an 'impenetrable fortress' to be built on the Airplot (Taylor, 2010), and Plane Stupid protesters threw a pie at business secretary Lord Mandelson after reports that he met with the head of BAA's public relations firm. Plane Stupid also 'hijacked' a table reserved for Virgin Atlantic Airways at a public relations awards ceremony.

By this point, BAA had announced that it would concede efforts to build a third runway if the Conservative Party formed the next government, which polls had shown would be a likely outcome of the elections. Prior to the elections, businesses attempted to pressure the Conservative Party to support the third runway, but the Conservatives would not budge. In the meantime, the Labour government continued with its support for expansion, and the House of Commons transport committee approved the third runway.

Before the elections, campaigners won a legal victory following a court challenge claiming that the consultation process was flawed, that expansion could jeopardize meeting climate change targets, and that there was not enough evidence to indicate that the government intended to provide necessary public transit to reach the runway. This challenge was initiated with the help of the documents obtained earlier by Justine Greening (Personal Correspondence, 17 September 2014), and it concluded with the High Court ruling that the government had failed to adequately consider the impact expansion would have on climate change considering already legislated emissions targets under the Climate Change Act. This did not result in a cessation of expansion plans but represented victory that, while important, soon became moot. When the elections came in May, no party won a majority of seats. The Labour Party was divided, and it did not take long before the

Conservative Party formed a coalition with the Liberal Democrats and opposed the third runway in their coalition manifesto as part of their energy and climate change platform.

On 24 May 2010, BAA formally dropped its intentions to build the runway. In October 2011, just after Ed Miliband assumed the position of Labour Party leader, Labour also took the third runway off its agenda. The coalition government did not completely reverse its position on airport expansion as a whole but, as Chancellor George Osborne stated, the government would 'explore all options for maintaining the UK's aviation hub status, *with the exception of a third runway at Heathrow*' (quoted in Stewart, 2012; emphasis mine).

Having been reassured of victory, the campaign settled down. However, a cabinet reshuffle in 2012 put some of the campaigning organizations on edge. Transport minister Justine Greening, a fierce opponent of the third runway, was moved to the position of Secretary of State for International Development. Patrick McLoughlin took her place, and the message moved from 'anywhere but Heathrow' to 'all options are on the table'. Some felt that a U-turn on the third runway was imminent, while McLoughlin attempted to calm them saying, 'I wasn't made transport secretary to push through [a] third runway at Heathrow' (as quoted in Murphy, 2012). At the same time, the government appointed Sir Howard Davies, former director general of the business lobbying organization Confederation of British Industry (CBI), to chair a commission on aviation expansion, which is set to be completed after the 2015 general election. This allows the Conservatives an opportunity to make a U-turn while fulfilling their coalition agreement with the Liberal Democrats.

In the meantime, the new cabinet published an aviation policy framework that removed previous strict thresholds for aircraft noise, and additional cries for a third runway at Heathrow came from the transport select committee, all while London mayor Boris Johnson called for a new airport to be built and for Heathrow to be closed down in order to develop housing in the area. Prior to the Davies Commission interim report, which was published on 17 December 2013, activists warned of new actions, a resurgence by groups such as Plane Stupid, and that '[t]housands of climate change protesters are on alert' (Blake, 2013). The interim report produced three options: a third runway at Heathrow, a doubling of Heathrow's existing runways, and a second runway at London Gatwick Airport. While Mayor Johnson felt that Gatwick would be expanded, Heathrow Airport Holdings (formerly BAA) established a new fund to insulate homes for noise and compensate homeowners in an attempt to pacify local opponents to the runway. It is uncertain if the third runway will be recommended by the commission, but it seems likely that if it is recommended, a resurgence of climate change activism and local campaigning will follow.

Campaigning for clean investment: the case of the Green Investment Bank

The idea for a green investment bank (GIB) came from Nick Mabey, director of the ENGO E3G, who wrote a report on the idea with Ingrid Holmes. This report was released in March 2009 and highlighted both the history of public banks as a means to further policy at the national and multinational level and listed the particular advantages a public bank has over limited ad hoc funding. It called the bank 'the best medium-term option for the UK' for 'delivering the next generation of low-carbon infrastructure' (Holmes and Mabey, 2009, 3–4). By this point, the Labour government had committed to the Climate Change Act. Based on the targets set by the act, over £160 billion in renewable investments by 2020 were required for centralized energy infrastructure alone (Holmes and Mabey, 2009). When the head of the economics team at FoE, Ed Matthew, read the report, he 'thought it was a great idea..., especially at a time of austerity.' He 'picked up the idea and used that paper to start lobbying the government on it' (Interview with Ed Matthew, 21 September 2012). The idea of the bank solved the problem of getting significant levels of finance from the private sector into green infrastructure in a practical manner (Interview with David Powell, 18 October 2012).

In order to push the government to deliver the idea, Matthew left FoE and established a new organization initially called Repower Britain and then renamed to Transform UK Alliance (or Transform UK) (Interview with Ed Matthew, 21 September 2012). Before doing so, however, he gauged interest for the idea by asking business and NGOs to sign a letter he wrote to *The Guardian* newspaper about the GIB. It received an 'absolutely positive' initial response; 'it really didn't take that much effort for the idea to really start to fly, so I thought it was an idea whose time had come' (Interview with Ed Matthew, 21 September 2012). The letter, calling on the bank to be included in the government's budget in 2009, appeared in *The Guardian* with nearly 60 signatures.

While the appeal to the Labour government failed, Matthew went on to form the alliance as the campaigning wing for the bank, bringing together a variety of organizations, including E3G, the environmental legal organization Client Earth, the environmental asset management group Climate Change Capital, the environmental business lobby Aldersgate Group, and others, including major ENGOs such as Friends of the Earth.

Aldersgate Group, a 'combination of a think tank and a lobby group' that speaks on behalf of its largely private company membership (Interview with Peter Young, Aldersgate Group chairman, 8 November 2012), played a major role in the alliance. Aldersgate, established in 2006, was particularly helpful as it had previously had access to the backdoors of ministries and Houses of Parliament and worked closely with professional and scientific bodies as well as ENGOs. The group was particularly interested in green investment. As their first task with the alliance, they set out to create dialogue between government and the financial sector. Aldersgate Group chairman Peter Young knew that communication between Westminster and the financial sector was 'very, very poor. Remarkably so' (Interview, 8 November 2012) and organized workshops to develop an understanding that if government wanted greater private investment in the green economy in order to achieve climate and other environmental targets, they would need to reduce private risk by making public investment.[20] This was the cornerstone of the GIB.

While Aldersgate continued to engage policymakers and the business community, Matthew moved the campaign to its next task: convincing the three largest political parties to commit to the green investment bank in their manifestos for the 2010 elections. With such a strong alliance, lobbying experience (Interview with David Powell, 18 October 2012; Interview with Peter Young, 8 November 2012), and ample political opportunities (see Chapter 5), the bank was an easy sell. Quickly, all three major political parties included the GIB in their manifestoes, to one degree or another. Unlike the Labour and Conservative Parties, the Liberal Democrats included a broader infrastructure bank that was not explicitly 'green' in their manifesto. The fact that the Liberal Democrats had the greenest policy agenda of the three parties (Friends of the Earth, 2010a) secured confidence in the campaigners that the bank would be environmentally sound (Interview with Ed Matthew, 21 September 2012). The quick success prompted the ENGO Green Alliance to state, 'the good news is that whichever party wins the next election, a Green Investment Bank will be established, with a clear low-carbon mandate' (as quoted in Hewitt, 2012).

The Conservatives moved particularly quickly on the issue. In late 2009, shadow chancellor George Osborne called on Bob Wigley, chairman of Yell Group plc., to head the Green Investment Bank Commission which formed in February 2010. Prior to the May elections, Labour government chancellor Alistair Darling announced that the 2010 budget would include the GIB with a modest investment of two billion pounds of equity, half from asset sales and the other half from private investment. This received a mixed response from campaigners, who were happy to see its inclusion but called for additional public investment. The campaign called for the Queen's Speech in May, regardless as to which party would win the election, to include not only an increase in investment but a 'Shadow GIB' that would initially focus on energy efficiency (Holmes and Mabey, 2010).

The elections resulted in a Conservative-Liberal Democrat government whose coalition agreement, after significant lobbying from campaigners, included an explicitly green investment bank (Interview with Ed Matthew, 21 September 2012). The Queen's Speech, however, was less positive. While it included a 'green deal' on energy efficiency for homes and businesses, it only suggested that the bill 'may' contain a GIB. Soon after, the GIB Commissions' first report was released stating that the scale of investment in renewables necessary to achieve climate change and renewable energy targets was £550 billion by 2020. The report also stated that the bank could be financed by consolidating existing funds and quasi-autonomous non-governmental organizations (quangos), which receive money from the government but are not under direct control of government or publicly elected officials. This would reduce inefficiencies of duplication and simplify the process for grant-seekers. Further funding could be found in providing long-term green bonds suitable for insurance and pension funds as well as green ISAs, or individual savings accounts. Importantly, the report called for the bank to be 'established by an act of Parliament as a permanent institution working over the long term in the national interest' (Green Investment Bank Commission, 2010).

It was not surprising that ENGOs responded positively to the report as many, including Green Alliance, Greenpeace UK, WWF UK, Climate Change Capital, and E3G either contributed or were members of the commission or the advisory panel themselves. Campaigners also formed a rough consensus on what features the GIB should have:

- Establishment in legislation
- Initial capitalization of £4–6 billion
- Borrowing powers
- Green ISAs and bonds

They felt assured that the coalition government would eventually deliver a GIB, but they campaigned for these components to be included. This campaigning was often done at the elite level and through backdoor lobbying. Even some public events were aimed at the green and financial sectors rather than the general public. Green Alliance held a debate on the GIB with speakers Bob Wigley, James Stewart of Infrastructure UK, Peter Young of Aldersgate, Tom Delay of Carbon Trust, and Nick Mabey of E3G. Also, Transform UK called on private organizations (for example, Microsoft, Bank of America, Merrill Lynch, BT, British Airways) and large NGOs (for example, WWF, E3G, FairPensions, Greenpeace) to sign a declaration calling on these features to be included following the commission's report.

Client Earth, hired by Transform UK, focused on producing a draft bill which they felt would demonstrate best practice at the same time as 'doing some of the thinking for the government' (Interview with David Holyoake, Client Earth, 25 October 2012).[21] The draft legislation took nine months to flesh out, starting well ahead of the government which, by early autumn, was divided on the issue as the Treasury was opposed to the GIB. Specifically, the Treasury was hesitant to legislate a bank with greater powers than existing quangos. However, campaigners' goals were supported by DECC, the Department of Business, and the Cabinet Office.

The divided government was pressured by the campaign to press the Treasury on the issue. Writing in *The Guardian*, founding director of E3G, Tom Burke, stated:

> The political reality is this. Britain will have a GIB if the prime minister really wants one. If he does not, we will have a Treasury designed label occupying the space where a real bank should be. We will also have answered two very big questions: the seriousness of the prime minister's claim to be green and whether ministers or Treasury officials are really running the country (Burke, 2010).

The Treasury, and especially Chancellor George Osborne, became the focus of the campaign, and although not many public actions were organized, Greenpeace UK took it upon itself to send activists to scale the Treasury building and display a banner that read 'Remember George – Green bank = New Jobs', helping to generate some media attention. Greenpeace also published and advertised a list of ten 'green quotes' by Osborne as a means to shame him. The quotes included 'Our commitment to the environment is as strong as ever', and 'Instead of the Treasury blocking green reform, I want a Conservative Treasury to lead the development of the low-carbon economy and finance a green recovery' (see Greenpeace, 2010a).

The Spending Review 2010 did not appear positive for campaigners. While it included the GIB, it only committed one billion pounds of funding and 'additional significant proceeds from asset sales...subject to a final design which meets the tests of effectiveness, affordability, and transparency' (HM Treasury, 2010, 62). One Whitehall source stated that 'The budget was £2bn at breakfast time on Tuesday, but only £1bn by lunchtime' (quoted in Vidal and Webb, 2010). While FoE remained positive, other ENGOs were disappointed by the news, citing a report commissioned by Green Alliance, Transform UK, and E3G which stated that the bank should be capitalized with £4–6 billion until 2015 in order to generate £450 billion needed in energy investment over the next 15 years (see Ernst & Young, 2010). Another crucial component, borrowing powers, was excluded from the review due to Treasury concerns that it would increase the national debt at a time of economic recession.

The Liberal Democrat climate change secretary, Chris Huhne, was reported to have cryptically criticized the Treasury to financiers over the issue of the bank and the lack of investment (Webb and Carrington, 2010), but it was not long before he fell more in line with the Treasury. Others also toed the Treasury line, leading campaigners to warn that '[b]-acktracking on plans to set up a green investment bank would not only renege on the coalition agreement, it would also seriously undermine David Cameron's pledge to be the greenest government ever' (Simon Bullock, FoE, as quoted in *The Guardian*, 2010).

In February, Transform UK ran an advertisement in *The Guardian* addressed to the prime minister, welcoming commitment to the GIB but calling on him to strengthen it to include the key features they demanded. It appeared to fall on deaf ears as, instead, the coalition government threatened to remove green ISAs from the bank despite investment industry support. They argued that government-supported green ISAs would create unfair competition with the small number of private green ISAs. Greenpeace, having listed the GIB as one of its campaigns for 2011, argued that ideology, rather than 'serious ambition to drive green jobs and growth' was driving the Treasury (in Harvey and Carrington, 2011).

Other ENGOs also attacked the Treasury, and an Environmental Audit Committee (EAC) inquiry report on the GIB sided with the ENGOs:

> The overwhelming majority of our witnesses supported the Green Investment Bank being a 'bank', able to raise its own finance, and not just another 'fund' to disburse government funding... It is clear to us from our many witnesses that the extent to which the Green

> Investment Bank is a 'bank' or a 'fund' is a key consideration as to whether the significant investment needed for the UK to meet its emission reduction and renewable energy targets will be raised. *We welcome the Business Secretary's ambition for the Green Investment Bank to be "a lot more than a fund", being able to lend and borrow. We recommend that Ministers deliver swiftly, and in full, on this ambition* (Environmental Audit Committee, 2011, emphasis in original).

The Treasury's main concern with granting the GIB borrowing powers, allowing it to be a fully-fledged bank, was that it would cause the size of government liability and debt to increase, at least on paper. This would be the case if the GIB would qualify as a public rather than a private institution. As the structure of the bank was undecided, there were some options.

The decision as to which institutions fall into the public accounts is made by the Office of National Statistics (ONS), who examine the institution's directorate, the independence of the board, funding sources, and other characteristics. If the bank was structured in such a way that the ONS would classify it as a private-sector institution, it would not affect the public sector net debt (PSND) or public sector net borrowing (PSNB) statistics beyond initial equity investment which would anyway be a transfer of funding from existing trusts and quangos. In addition, precedent had been set by other institutions that were temporarily placed off the public books, such as part-publicly owned Lloyds Banking Group and the Royal Bank of Scotland. In the case of the GIB, ministers had not even inquired into the specificities of the ONS classification to attempt to allow the GIB greater powers without increasing the PSND and PSNB.

The EAC report recommended that if the government was unable to remove the GIB from the PSND statistics, they should develop a private financial institution and keep the bank off the public balance sheet. Campaigners were conflicted by this recommendation. Green Alliance and Transform UK were wary of the government curtailing powers of a bank that was on the public books but understood that a bank not backed by the government would lose a key characteristic for raising capital. These organizations chose to support a public bank with curtailed powers in the hopes that it would regain powers and be taken off the public books in the long term. Not long after the EAC report, the government made its decision on borrowing powers for the bank: they would not be included until at least 2014, pending the decision of the next spending review. The bank had, in effect, been turned into a fund. However, the Treasury stated that the GIB would classify as a public institution.

The campaign did not give up hope as the bank was not yet written up as legislation, so amendments could still be made, and it seemed as though lobbying was able to make some headway. In March 2011, Osborne announced an additional two billion pounds for the bank from the sale of government assets while also suggesting that the GIB could be up and running by 2012. He also reiterated that the GIB would not be given immediate borrowing powers but that these powers would be granted in 2015, effectively only just delaying the delivery of a fully-fledged bank. Many campaigners were disappointed that borrowing powers would not be immediate, but the GIB Commission chair Wigley and Energy and Climate Change Select Committee chair Tim Yeo were both pleased with the compromise. One insider stated that the delay in borrowing powers would have little effect because key investments would be in offshore wind projects that would unlikely need investment before 2015 (see Harvey, 2011). However, Labour Party leader Ed Miliband and the chair of the EAC both called for immediate borrowing powers. Their message was echoed by Transform UK.

The details of the GIB were announced in May by deputy prime minister Nick Clegg indicating that the bank would indeed be legislated but would only have borrowing powers starting in April 2015 'on the basis...that the government target for debt to be falling as a percentage of gross domestic product has been met' (Clegg, 2011). Business secretary Vince Cable, a proponent of an infrastructure bank that was not necessarily 'green', later revealed detailed plans, including allowing the bank to fund flood defenses and nuclear power. He also suggested that Clegg's original promise of an independently borrowing bank would be limited to borrowing through the Treasury. It was later reported that Cable had attempted to keep both Lord Nicholas Stern, who authored the Stern Review on the Economics of Climate Change, and Bob Wigley from being on the advisory board of the GIB, both of whom were later granted those seats by DECC, indicating further tensions between the ministries.

Then in December, the bank faced another assault. A leaked draft report by the Treasury and the Department of Business, Innovation, and Skills (BIS) indicated the following:

1 The bank's funding would reduce from three billion pounds to 'up to £3bn'.
2 The bank would be instructed to make an annual profit, removing the option to fund more risky and experimental investment projects.

3 The bank could not invest more than 5 per cent of its funds into a single scheme, meaning that it could not play a leading role in some development projects.
4 The government was considering setting strategic priorities threatening the independence of the bank's investment choices.

The strategic priorities of the GIB proved a controversial issue for the government, with DECC calling for the funding of fledgling innovations and BIS pushing for funding proven technologies. BIS won the battle when the strategic priorities were limited to offshore wind power, commercial and industrial waste processing and recycling, energy from waste generation, and domestic and non-domestic energy efficiency.

Campaigners received another blow when Labour's shadow energy and climate secretary Caroline Flint noted that, according to figures that showed the government was set to borrow more money than planned, it was unlikely the debt target would be met in 2015, which could delay the granting of borrowing powers for the GIB. Rather than strengthening the GIB, the next government announcement was that the location of the bank's headquarters would be split between Edinburgh and London, to which FoE's economics campaigner David Powell said, 'Choosing the [headquarters] for the green investment bank has been like arguing about where to put the cherry on a half-baked cake' (quoted in Friends of the Earth, 2012c).

On 14 May 2012, the Enterprise and Regulatory Reform Bill (ERRB), which included the GIB, was introduced to Parliament following its mention in the Queen's Speech. Shortly after, Client Earth drew up a statement regarding the provisions on the GIB raising three concerns:

1 The bill failed to adequately restrict the bank from funding high-carbon projects. There was some concern here that the bank could turn into something other than a 'green' bank, although Client Earth and Aldersgate were not concerned this would occur in the immediate future (Interview with Peter Young, 8 November 2012).
2 Client Earth claimed that the legislation 'contains no indication that the bank will be ever be allowed to borrow in practice'.
3 They stressed a lack of transparency and accountability (Client Earth and Transform UK, 2012).

Client Earth proceeded to draft amendments to strengthen the bank and discussed these amendments with Labour MP Iain Wright, who presented the amendments, as written by Client Earth, in the House of Commons (Interview with David Holyoake, 25 October 2012).

In June, Client Earth and Transform UK submitted a memorandum on amendments to the ERRB. Their first suggested amendment was to allow the GIB to 'borrow from the capital markets no later than June 2015' (Transform UK and Client Earth, 2012). The memorandum also called on an independent board to review the bank's performance every five years and to allow public scrutiny because otherwise, 'the government shareholder will have the ability to amend the bank's priority sectors at whim' (Transform UK and Client Earth, 2012; also see Client Earth, 2011). The memorandum stated that the initial legislation is unclear about which projects could be funded by the GIB, therefore 'providing no legal certainty that the bank will be focused on unlocking investments in transformational green technologies' (Transform UK and Client Earth, 2012). In order to ensure green investment, they recommended amending 'The Green Purposes' clause of the bill from reading 'The advancement of efficiency in the use of natural resources' to 'Accelerating significant improvements in energy savings and energy efficiency' (Transform UK and Client Earth, 2012). In addition, they specifically added a clause mentioning that the aims should be pursued in line with the Climate Change Act 2008 emissions targets, and for good reason:

> The Climate Change Act is internationally known. It's our bit of flagship policy. It's been accepted by all three political parties. ... So I feel that it has the most longevity and durability of the options that have come up. So why invent something else which I don't think has the same durability as the Climate Change Act? (Interview with Peter Young, 8 November 2012)

Around this time, FoE launched an online initiative calling on supporters to lobby their MPs to strengthen the GIB by granting borrowing powers and insuring its green purposes. Local FoE groups held campaign events on the issue and ENGOs signed a letter to Prime Minister Cameron and Deputy Prime Minister Clegg stating that the GIB would fail without being able to borrow.

A divide remained within the government. Campaigners felt that Clegg had become an important ally (Interview with Peter Young, 8 November 2012), and Cable had reassured them that borrowing powers would be granted to the GIB no later than 2015/16. However, Cable's desire for a broader infrastructure bank put these comments under a cloud of suspicion (Interview with Ed Matthew, 21 September 2012). Campaigners remained hopeful and continued to lobby the government as the legislation made its way through Parliament.

The central concerns remained the power to borrow and the strength of the bank's green purposes. Ed Matthew felt strongly that the bank would remain green but felt less certain that it would be given borrowing powers (Interview with Ed Matthew, 21 September 2012). Peter Young of Aldersgate felt more hopeful (Interview with Peter Young, 8 November 2012). A bit more hope appeared when delegates at the Liberal Democrat Party conference approved a motion to make immediate borrowing powers for the GIB an official party policy, increasing coalition tensions. FoE urged the Liberal Democrat chief secretary to the Treasury, Danny Alexander, to push Osborne on the issue. However, at the same conference, Vince Cable reiterated his desire for a broad infrastructure bank, again threatening the GIB's green purposes.

The campaign's amendments had largely failed as the legislation went through the House of Commons. Insufficient time was allocated to discussing the GIB[22] (see House of Commons Hansard Parliamentary Debates, 2012, c.361–90), and Labour MP Iain Wright's amendment to strengthen the 'green purposes' clause lost (222 to 285), while the government's amendment that left the door open for borrowing powers to be further delayed was passed. Campaigners were disappointed but were informed that the government may introduce an amendment regarding the green purposes clause. However, campaigners remained skeptical:

> Obviously we have all been in positions before I think as lobbyists when the government has promised to go away and introduce its own amendment and either they haven't done it or the amendment that's put forward is weak, full of holes, and doesn't do the job (Interview with David Powell, 18 October 2012).

Lobbying continued as the legislation entered the House of Lords, where campaigners felt they had considerable support (Interview with Peter Young, 8 November 2012; Interview with David Holyoake, 25 October 2012). Despite the enthusiasm, debates in the Lords proved equally difficult for opposition amendments. Initial discussion looked hopeful, with Labour Lord Stevenson of Balmacara and Liberal Democrat Lord Razzall supporting borrowing powers and increased funding. However, Lord Smith of Kelvin, who had been appointed as chair of the GIB, suggested that the bank could manage high standards of greenness itself, without it being legislated, and that borrowing powers were not needed immediately. Regarding the latter point, Kelvin argued that the bank needed to 'show government and private capital markets that we are a well-run organization with a good track record worthy of the injection of more capital or, indeed, borrowing money in capital markets' (House of Lords Hansard Parliamentary Debates, 2012, c.1529). In addition, he stated that, 'if we feel we need to borrow we will approach the [government] shareholder well before 2015' but that he was 'confident that we can commit £3 billion wisely by 2015' (House of Lords Hansard Parliamentary Debates, 2012, c.1529). Kelvin's statements proved useful for the government, who repeated them when countering amendments brought by former FoE climate change campaigner Baroness Bryony Worthington, Lord Teverson, and others covering a range of issues. However, the government did pass an amendment regarding the GIB's green purposes so that, as a whole, the bank's investments are 'likely to contribute to a reduction of global greenhouse gas emissions' (Enterprise and Regulatory Reform Bill, 2013, 8).

The GIB opened on 28 November 2012, being cautiously celebrated by ENGOs. The bank made its investments in energy efficiency, a biomass power plant, offshore wind turbines, a waste treatment and recycling plant, anaerobic digestion plants, combined heat and power units, and renewable energy boilers. As of 31 March 2014, the GIB has invested in 31 projects, committing £1.3 billion. While the campaign was able to achieve their primary objectives, it remains to be seen whether the GIB will be given borrowing powers, allowing it to help mobilize the levels of investment into renewable energy and energy efficiency needed to meet the UK's climate change targets. Without those powers and additional government financial support, private investors remain wary of investing in green projects. The lack of financing threatens the transition to a green, low-carbon economy.

Conclusion

The case histories above are examples of national policies that were enacted to mitigate climate change. The Climate Change Act created a long-term emissions target. While the government was initially hesitant to include a strong target of 80 per cent reductions by 2050, this target was eventually adopted. The Green Investment Bank was also legislated, although pressure to make it stronger failed. Its future is still uncertain as borrowing powers have not been granted, meaning that it remains nothing more than a green investment fund. In the case of Heathrow's third runway, the Conservative Party U-turned on the issue. They and the Liberal Democrats stopped the runway after they formed a coalition government in 2010.

These cases are also examples of campaigning efforts by the climate change movement. FoE initiated the Big Ask campaign that called for the climate change act. These efforts were supported by their local FoE chapters as well as Stop Climate Chaos. Protests were organized and events were held. Celebrities were mobilized and MPs were lobbied by their constituency. Arguments were developed and the public was kept informed.

Although the campaign against the expansion of Heathrow was initiated by local campaigners, climate change activists soon joined them in a multifaceted attempt to stop a third runway from being approved and built. Local campaigners informed nearby residents of new developments and called on them to oppose the runway. Plane Stupid coordinated widespread direct action. Greenpeace climbed atop planes and organized the Airplot. HACAN and NoTRAG commissioned reports and participated in protests. Flash mobs appeared at Heathrow Airport and local councilors lobbied their party leaders.

The Green Investment Bank was an idea that sprouted from an ENGO, and a campaign was organized around it. At its heart was Transform UK, but others, including Aldersgate Group and Friends of the Earth, played important roles. Transform UK headed the lobbying efforts, speaking with policymakers and convincing them to take up the idea. Aldersgate Group mobilized business group support for the bank. Client Earth helped draft amendments to the government's bill while Greenpeace displayed a banner across the Treasury building calling for a strong GIB.

Though we see in these cases that climate change policy developed in the UK, and that campaigning took place, was the policy change a result of these campaigning efforts? If so, to what extent? Which part of the policy process was actually affected by the campaigns? In short, what impact did the campaigns have on these policies? This is the subject of the next chapter.

4
Policy Outcomes

To understand social movement outcomes, we investigate *what* those outcomes are. Saying that a campaign preceded policy change is not enough. We must show that the movement played a role in that policy change. Using the case of the climate change movement in the UK, specifically looking at the campaigns to create the Climate Change Act, stop the third runway at Heathrow, and establish the Green Investment Bank, I will explore which outcomes the campaigns were able to achieve. I do this by applying a counterfactual approach. A counterfactual approach asks 'what if' questions and uses data to provide a solid answer. Here we are asking the question, What would have happened to the policies if there were no social movement campaigns?

For the purposes of this book, I am interested in movement impact of policy formulation and policy legitimation within the policy cycle (Kraft and Vig, 2006). 'Policy formulation' refers to the ways in which governments establish goals, develop options, and deal with policy problems; 'policy legitimation' concerns the passage and adoption of policies and their effectiveness in being integrated into further policymaking considerations. In other words, policy formulation regards what policy is developed, and policy legitimation is the strength and seriousness of the adopted policy.

In order to investigate impacts, I have developed a model to better examine and understand the changes caused by social movements. It is worth remembering that early conceptualizations of movement outcomes focused on success and failure, but this binary approach does not consider the whole scope of changes that occur in the direction desired by the campaigners. In most cases, campaigners will state a specific desired goal but would rather have something akin to that goal than the status quo. Therefore, the research presented here looks to gauge the influence of campaigners on the *direction* of the campaign's goals (see Table 4.1).

Table 4.1 Direction and desirability of movement outcomes

Campaign	Less desired	⟵	⟶	More desired
Climate Change Act	No legislated targets	Legislated but weak targets	Legislated strong targets with weak accountability	Legislated strong, accountable targets including international aviation and shipping
Heathrow Third Runway	Third runway approved	Third runway with significantly capped flights	Third runway with significantly capped flights and strong environmental standards	Third runway rejected
Green Investment Bank	No Green Investment Bank	Green Investment Bank without borrowing powers and with limited government investment	Green Investment Bank with borrowing powers but with limited government investment	Green Investment Bank with borrowing powers and significant government investment

The model I developed divides up policy outcomes into several components, and I will examine each component in each campaign. It focuses on the movement's influence on setting policy options, obtaining and expanding political support for its cause, influencing the actions of policymakers, and making the desired change in policy either through the creation of legislation approved by the campaigners or a change in policy positions by government. By using the model to gauge the strength of outcomes in the direction of desired change for a wide variety of outcome components, we can circumvent the problems posed when relying on crude quantitative indicators (for example, the number of bills passed or government spending) used in previous research.

The Policy Outcomes Model (see Table 4.2) encompasses policy consideration, political support, political action, desired change, and desired outcome within a particular campaign or policy change.

Policy consideration is a more specified version of agenda setting. 'Agenda setting' refers to the establishment of 'the list of subjects or problems to which government officials, and people outside of government closely associated with those officials, are paying some serious attention at any given time' (Kingdon, 1995, 3).[23] Instead, policy consideration refers to examining the breadth and depth to which the campaigners problematize or advocate a *particular* policy and make it politically salient. For example, rather than setting the agenda on mitigating climate change, the campaign sets an agenda on legislating specific emissions targets. This policy consideration component begs such questions as the following: How quickly did the campaign grab hold of policymakers' attention on the issue? How long was this attention held? Which policymakers are considering the issue, and how much power do they have regarding that decision? How salient did the campaign make the issue for policymakers relative to other concerns? By answering these questions, we can understand the impact campaigners had in this area.

Table 4.2 Policy outcomes model

1	**Policy Consideration**
	Speed of gaining policymakers' attention
	Length of attention held by policymakers
	Level of political hierarchy where discussion takes place
	Salience relative to other concurrent issues
2	**Political Support**
	Number of supporters
	Strength of support
	Level in political hierarchy of supporters
	Support from gatekeepers
3	**Political Action**
	Political strength of the action
	Actors' level in political hierarchy
4	**Desired Change**
	Secureness of policy change
	Strength of disincentives for not fulfilling policy's mandate
	Breadth and depth of policy's mandate
	Strength of implementation
5	**Desired Outcome**
	Strength of enforcement
	Level of scale diffusion
	Level of space diffusion
	Level of institutional internalization

Political support examines the extent to which a movement or campaign was able to attract political allies to its policy position and develop political traction. This can help assess the amount of pressure the campaign can put on policymakers, even without the passage of policy. The level of political support can go on to be an important factor in the direction of the campaigners' policy aims even if they fail to achieve their primary objective. In order to understand the importance of the campaign in attracting political support, I seek to answer the following questions: How many supporters were the campaigners able to attract to their policy position? Was the support by these policymakers strong or weak? How powerful were the political supporters of the campaign's policy position with regard to the policy? Was the campaign able to acquire the support of gatekeepers or key policymakers with some amount of veto power (Busby, 2010)?

Political action refers to direct attempts at policy change through legitimate political channels in the policy direction desired by campaigners. This can include the sponsorship of bills, citizens' initiatives such as a referendum, or a court challenge. Although legislation may not successfully be enacted, the extent to which campaigners were able to influence political action is worthy of exploration. In order to gauge the influence of campaigners, we should answer these questions: How powerful or serious are the political actions that are taken? Who is taking the action, and how much power do they have over the policy area?

Desired change is the extent to which the policy change is formulated and functions in a direction favored by the campaigners. This is where the campaign's influence on the particularities of a policy change is assessed. These particularities include the secureness of the policy change, the strength of the disincentives for not fulfilling the policy's mandate, the breadth and depth of the policy's mandate relative to the desires of the campaign, and the strength of implementation suggested in the policy. It is necessary to gauge the levels of secureness, disincentives, breadth and depth, and implementation strength in order to get a comprehensive understanding of the campaign's influence on desired change.

One important component, *desired outcome*, is missing from this analysis. Desired outcome refers to the campaign's influence in achieving the desired 'policy direction objectives' following political action. For example, since the Climate Change Act calls for an 80 percent reduction of greenhouse gases by 2050, one way to know if the desired outcome has been achieved is to see whether the policy was able to produce that reduction and what role the campaign had on its successfulness. Examining the desired outcome is outside the scope of this book because it requires a long passage of time following the policy change. Obtaining a measure of desired outcomes would require answers to several questions: What was the movement's role in the strength of enforcement that the policy was granted? To what level did the movement contribute to the diffusion of the policy between political institutions across scale (for example, across local, regional, and national government) and space (for example, across countries)? What level of contribution did movements make to political institutions' internalization of the policy change?

Below I examine each campaign and analyze the role of campaigners in achieving policy outcomes in each of the components of the model.

Climate Change Act

Prior to the Big Ask campaign, the Labour government had pledged to reduce greenhouse gas emissions 60 percent by 2050 on the recommendation of the Royal Commission on Environmental Pollution. Both the Labour and Conservative parties had included those targets in their manifestos, although they made no promises to pass legislation mandating these targets.

FoE, wanting to secure the pledged reductions, began to push for legislation calling on emissions reductions of 3 percent annually. FoE started to influence *policy consideration* by meeting with key policymakers outside of the cabinet, including two former environment ministers and the Liberal Democrat environment spokesperson. These were not gatekeepers on the issue but important, cross-party voices in the policy area. They were quick to hear out and agree to FoE's idea, having been in regular contact with FoE prior to this particular campaign.

The government did not quickly adopt the legislation. Instead, having developed policy consideration across parties, the campaign was able to attract the attention of the opposition leadership, which increased the issue's salience enough to create political competition around the policy. Despite not having their policy adopted by the government initially, the campaign was able to have the ear of government for some time. This suggests that the Big Ask campaign was very important in the policy consideration for the Climate Change Act.

Concerning *political support,* it was clear that once the policy was considered, the initial supporters of the presentation bill were quick to endorse the campaigners' position. Backbench and opposition MPs then proposed an early day motion, following the lead of the campaign (*political action*), as a way to pressure the government to take on the legislation. This coincided with a media blitz by FoE's Tony Juniper and Radiohead's Thom Yorke. By the time the backbench/opposition bill was introduced to Parliament, EDM 178 had been signed by over 200 MPs, including 44 Conservative, 45 Liberal Democrat, and 108 Labour MPs.[24] The ease with which MPs signed the EDM suggests that 'FoE was pushing at a door that was already at least half open' (Rootes, 2011, 61) and that the campaign cannot be credited with achieving this support.

Figure 4.1 shows that the early support for the EDM tapered off in a few months, while it took another year for the bill to be adopted by the government. This suggests that the initial wave of support may not have been enough to pressure the government to adopt the bill, a claim that was backed by correspondence with several MPs (Jim Sheridan MP, 27 March 2012; Geoffrey Robinson MP, 26 March 2012). It was fairly clear that this first wave was aligned with the policy position without needing further pressure from the campaign. FoE continued to campaign to attract more support from other MPs:

> Our tactic of targeting MPs who haven't seen climate change as a priority issue, and ensuring that they get as many postcards, letters and emails as possible from their constituents in support of the bill is really working. Getting 400 MPs to sign the EDM by the autumn is a real possibility (Friends of the Earth Hammersmith and Fulham, 2007).

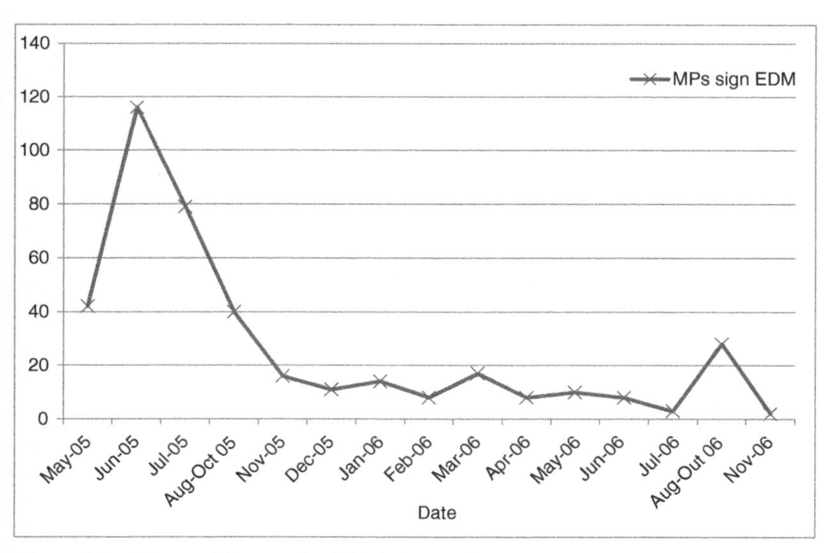

Figure 4.1 Climate Change Act EDM signatories

Was it the campaign that attracted additional support for the EDM following the initial wave of signatories? In order to test this, I examined both the influence of campaigning actions and the media coverage of these actions. Following the initial wave of support, the campaign consisted of a wide variety of actions in order to solicit MP support for the Climate Change Bill and for the EDM in particular. This included the Carbon Speed Dating Lobby and other lobbying events, particularly during the Big Month of lobbying starting in September 2006, which Tony Juniper felt was important for pressuring the Labour Party into taking on a climate change bill (Interview, 18 September 2014). These other events included concerts promoting the campaign, public demonstrations, the publication of relevant reports, and press conferences. One way to see if these events had an effect on MP support is to see if they coincided with increases to the number of EDM signatories.

Using extensive documentary analysis of campaign-affiliated literature and newspapers data, I constructed a list of all campaign actions following the first wave of EDM signatories and noted their dates as accurately as possible. I coded the various actions into three categories of campaigning: 1) public events, such as protests or concerts, 2) lobbying events, where the main purpose of the event was aimed at lobbying MPs, and 3) the publication of research used to argue the case for the act. These actions started in March 2006 when the campaign reignited.

In addition to the actual campaign actions, I also looked at news coverage of the Big Ask campaign and compared it with EDM signatories. News media can both influence public opinion (Page *et al*, 1987) and inform politicians of issues and levels of public concern (McCombs and Shaw, 1972). In this way, events that appear in the news can work to increase the importance of postcards MPs receive or other lobbying attempts. By looking at newspaper articles on the campaign efforts and on the proposed legislation, I test media attention's effects on the actions of politicians. I collected data from UK national newspapers using the electronic database Nexis and ran two searches looking at national newspaper coverage. First, I looked for 'The Big Ask campaign' or 'Stop Climate Chaos' or 'climate change act' or 'climate change bill'. The second search was: 'The Big Ask' and 'Friends of the Earth' or 'FoE' or 'Juniper' or 'Thom Yorke'. Both searches looked at articles between May 2005, when campaigning began, and 14 November 2006, when the Climate Change Bill was announced in the Queen's Speech. I then manually examined the articles, filtering out those that were unrelated.

Media coverage was high in May 2005 when the campaign began but then dipped in November. Media coverage increased in March 2006 when public campaigning reignited and Radiohead's concert for the Big Ask Live led to some media coverage in May. June, July, and August represented months with no large actions taking place, so no news articles were published, except for a June article on Thom Yorke in the music section of *The Observer* that included substantial discussion on the Big Ask campaign. News coverage increased in early September after David Cameron met with Tony Juniper to give support to the campaign and later with the beginning of the Big Month lobby. Media attention again grew just before the announcement of the bill in the Queen's Speech, peaking in October (see Figure 4.2).

To better understand the relationship between news articles and signing the EDM, I aggregated the independent and dependent variables over a seven-day period in order to lag the data, comparing the independent variables to the dependent variables seven days later. Findings from previous studies suggested that events and news coverage continued to influence policymakers anywhere from two to four weeks after they occurred (Wood and Peake, 1998; Bartels, 1996; Walgrave *et al*, 2008). However, the relatively easy task of signing an EDM as compared with the measures used in those studies (which included speeches and congressional hearings) made one week a better time period for comparison. Therefore, I compared newspaper articles and campaign actions each aggregated over seven days starting from 1 March 2006 with the aggregate number of MPs who had signed the EDM in the following seven days. For Parliamentary recesses, when MPs would be unable to sign the EDM, I aggregated newspaper article and campaign action data, respectively, over the days of the recess and any days prior to the recess following the previous weekly aggregate. I then compared this data with the aggregate of MPs who signed the EDM in the seven days following the recess.[25]

I utilized a statistical analysis known as Kendall's tau to measure the association between the actions and newspaper coverage with the number of MPs who signed the EDM one week following the action. Kendall's tau is used since the number of actions is relatively small for a statistical test and the nature of the variable requires a nonparametric statistic (Ferguson *et al*, 2011; Helsel and Hirsch, 1992, 212).

The association between news coverage of the campaign and EDM signatories, and between campaign lobbying efforts and EDM signatories, appear significant (see Table 4.3). Newspaper articles associated with MP signatories in the week following their publication appear at the .05 level

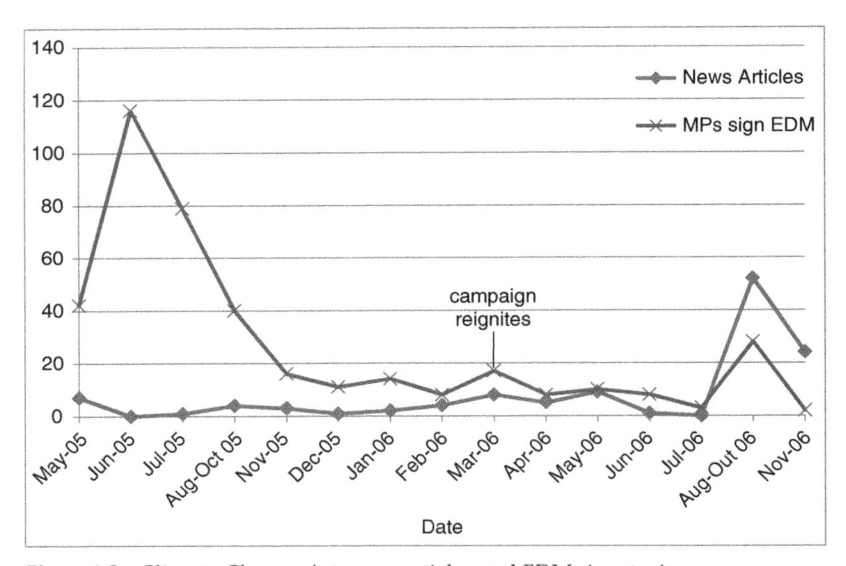

Figure 4.2 Climate Change Act news articles and EDM signatories

Source: Nexis 'The Big Ask campaign' or 'Stop Climate Chaos' or 'climate change act' or 'climate change bill'; 'The Big Ask' and 'Friends of the Earth' or 'FoE' or 'Juniper' or 'Thom Yorke'; EDM 178, 2005.

($\tau = .409$). This suggests that newspaper reports on the act had an impact on MP signatories and that the effects of campaigning events may have been mediated through newspaper coverage. Campaign actions also had significant associations with MP signatories. Total actions and MPs signing the EDM associated positively and significantly at the .10 level ($\tau = .347$). Interestingly, lobbying was the only form of campaign action with a significant and positive result ($\tau = .527$, $<.1$). This suggests that the campaign played a role in garnering additional support following the initial wave of signatories, even when the actions were not explicitly calls for MPs to sign the EDM. This was also backed by correspondences with MPs, some of whom stated that lobbying efforts played a role in their own decision to sign the EDM (Jim Sheridan, 27 March 2012; Alexander Stafford, Office of Andrew Rosindell MP, 3 April 2012). While other MPs did not state that lobbying played a role (Geoffrey Robinson MP, 26 March 2012; Peter Lilley MP, 25 March 2012), the evidence suggests that a significant number were still motivated by the campaign.

The campaign attempted to gain support beyond simply asking MPs to sign the EDM. They knew that in order for a strong bill to pass, even

			Total Articles	Public Actions	Lobbying	Reports	Total Actions	EDM Signatories
Kendall's tau	Total Articles	Correlation Coefficient	1.000					
		Sig. (2-tailed)	.					
		N	19					
	Public Actions	Correlation Coefficient	.423*	1.000				
		Sig. (2-tailed)	.044	.				
		N	19	19				
	Lobbying	Correlation Coefficient	.494*	−.028	1.000			
		Sig. (2-tailed)	.015	.902	.			
		N	19	19	19			
	Reports	Correlation Coefficient	.380	.327	.090	1.000		
		Sig. (2-tailed)	.071	.166	.695	.		
		N	19	19	19	19		
	Total Actions	Correlation Coefficient	.698**	.597**	.642**	.480*	1.000	
		Sig. (2-tailed)	.000	.006	.002	.028	.	
		N	19	19	19	19	19	
	EDM Signatories	Correlation Coefficient	.409*	.046	.527**	−.082	.347	1.000
		Sig. (2-tailed)	.028	.824	.009	.691	.073	.
		N	19	19	19	19	19	19

*Significant at the 0.05 level (2-tailed).
**Significant at the 0.01 level (2-tailed).

if the government had adopted the general policy, MPs would need to strongly support the campaign's agenda.

> If in three or four years' time these MPs come under pressure from their parties to vote against the bill, the thought that they once signed an EDM is unlikely to be enough to make them resist the whips. If on the other hand they have publicly supported the bill in lots of other ways, such as speaking at public meetings, working with local campaigners to publicise the bill or raising the issue in Parliament, then they will be much less likely to do a U-turn at the last minute (Friends of the Earth Birmingham, 2005).

One way to increase MP support and force the government to take on the bill was trying to commit MPs to writing letters to the government in favor of the bill (Interview with Tony Juniper, 18 September 2014; see for example, Friends of the Earth, 2006d). FoE made a call to do just that during the Big Month lobby (Friends of the Earth Ealing, 2006). This and other calls for support allowed for a variety of 'asks' the campaign could make of MPs, all of which would amount to supporting the bill. For example, Lewisham FoE held a 'Big Ask Public Meeting' where community members could listen to speakers and local MPs on the topic of the Climate Change Bill. Campaigners followed up by issuing press releases that included MPs' quotes from the event (Burke, 2006).

As part of the campaign strategy, FoE and others worked to get a cross-party consensus around the issue. Campaigners found it important to gain the approval of the shadow cabinet at around the same time as the government (see Carter, 2006). This mitigates concerns about the possible implications of elections shifting the balance of power (Interview with Tony Juniper, 18 September 2014) but also pressures the party in government to act. In this case, it was even more important to gain opposition party support because the Conservatives were seeking to change their image and become greener (Worthington, 2011; see Chapter 5).

Cameron's support for the Big Ask campaign increased pressure on the Labour Party to act on climate change (Carter and Jacobs, 2013, 9). While it may have been political maneuvering that resulted in the Conservative Party leadership supporting the campaign, the campaign itself was important for obtaining strong public support because it provided an independent and green cover for the Conservatives' decision. This would explain why David Cameron did not simply go on television to announce Conservative Party approval for the Climate Change Bill, but did so alongside FoE director Tony Juniper, relying on the popularity of and public trust in FoE. Cameron's outspoken support qualified as a small *political action* by taking a policy position on the bill. It was part of a series of moves made by the Conservative Party to green their image, and FoE provided them with an opportunity to do the same around the bill.

Additional political action was taken once the idea of the bill was supported by the government. First, the bill appeared in the Queen's Speech, partially as a result of campaigning and partially in response to the political party maneuvering (Interview with Norman Baker, 12 March 2012; Interview with John Gummer, 26 March 2012). Then the government drafted the legislation and called on Friends of the Earth campaigner Bryony Worthington to draft the bill. When the draft was in Parliament, FoE wrote amendments and recruited MPs and members of the House of Lords to propose them. Although the amendments failed, FoE played a part in the political action taken by those policymakers.

They were also able to influence *desired change*, to some extent. First, a legislated emissions target was vital as it secured the policy change considerably more so than pledges by the political parties. However, the campaign was unable to apply enough force to create the desired policy straight in the government's draft, nor ultimately decide the fate of their key amendments. However, there was some sign that campaigning was effective in strengthening the 2050 target to 80 percent. During a Parliamentary debate on the Climate Change Bill in which the House of Commons was discussing emissions targets, Labour MP Phil Woolas stated:

> I admire the Conservative Party's position in not jumping on the 80 per cent bandwagon. When we announced the policy change, I said, "If we go for 60 per cent, everyone else will go for 80 per cent." I am as sure as eggs is eggs that if I announced 80 per cent today, postcards galore would be flying into the office by Monday morning demanding 90 per cent (House of Commons Hansard Parliamentary Debates, 2008b, c.107).

The government deferred the decision on the emissions target until the shadow Committee on Climate Change's report. It turned out that Woolas was incorrect in his assessment as FoE who, like the committee, had relied on recent scientific data to justify their target recommendation. It must be said, however, that the committee could have been asked to inform the government on their recommendation following the passage of weaker legislation and following the committee's official establishment, rather than during its shadow stage. Additionally, there was no assurance that the government would abide by the recommendation

(House of Commons Hansard Parliamentary Debates, 2008a, c. 61; although see Carter and Jacobs, 2013, 12). It is likely that the campaign did influence early consultation of the committee and ensured that their recommendations would be addressed in legislation (Interview with Tony Juniper, 18 September 2014).

Regarding year-on-year targets, FoE was satisfied with a middle ground that the government's bill featured, because it captured the essence of 'a pathway' to emissions reductions that would lead to an 80 percent cut by 2050 (Interview with Tony Juniper, 18 September 2014). The inclusion of international aviation and shipping, FoE's other desired amendment, risked confrontation with Europe. FoE's campaign was not powerful enough to offset the 'concession costs', or the 'anticipated losses resulting from acceding to movement demands' (Luders, 2010, 3) on this issue. Nonetheless, the emissions levels are said to be accounted for in the present budgets. While the act removed the monetary penalty on ministers, the publicity around the Climate Change Act generated by the campaign makes it an act that would be difficult to U-turn on, although backbench Conservative MPs have designs on just that (see for example, Greenpeace, no date).

Table 4.4 Policy outcomes model – Climate Change Act

Policy Consideration	**Very Important**
Speed of gaining policymakers' attention	*****
Length of attention held by policymakers	****
Level of political hierarchy where discussion takes place	****
Salience relative to other concurrent issues	***
Political Support	**Moderately Important**
Number of supporters	**
Strength of support	***
Level in political hierarchy of supporters	**
Support from gatekeepers	*
Political Action	**Very Important**
Political strength of the action	****
Level in political hierarchy of actors	***
Desired Change	**Important**
Secureness of policy change	****
Strength of disincentives for not fulfilling policy's mandate	**
Breadth and depth of policy's mandate	***
Strength of implementation	**

*indicates level of importance the campaign had in attaining each sub-component. These are averaged to determine the level of importance of each component.

The Climate Change Act as a whole was influenced significantly by the campaign, which was particularly important in affecting policy consideration and political action (see Table 4.4). The campaign's impact was acknowledged by policymakers across political parties (see for example, Friends of the Earth, 2006e, 2006f; also see Friends of the Earth, 2007a), which may be particularly telling as the policymakers stood to lose by putting emphasis on the accomplishments of the campaign for something as significant as the Climate Change Act rather than praising their own efforts or that of their party.

Heathrow third runway

Many local residents were continuously opposed to expansion at Heathrow, some for decades (Interview with John McDonnell, 21 June 2012). While there were occasional public protests prior to the campaign opposing the third runway, many of the local dissenters tried to make their voices heard through the public consultation process. This certainly informed the public inquiry and in the case of Terminal 5 the sheer number of consultation documents and public comments led to a significant delay in the expansion process. Local policymakers were also vocal in their opposition and informed their party leadership of their concerns.

In the case of the third runway, the local public and local policymakers were opposed to expansion from the outset. Coming on the heels of Terminal 5 approval, the industry called for a third runway, and the government's willingness to accommodate the proposal quickly led to public displays of discontent, but often in the form of individual actions. The consolidation of two local groups following the failure to stop Terminal 5 was at the heart of collective opposition. They pushed policymakers to consider their policy position in various ways. Direct communication with government officials and party leadership came from local MPs who were members of the campaign themselves. Labour MP John McDonnell in particular played a vocal role in both speaking to the Labour government and engaging in campaign events. The same could be said of Zac Goldsmith for the Conservative Party, who also fundraised for the campaign prior to becoming an MP (Personal correspondence with Zac Goldsmith, 15 September 2014).

Another, perhaps more crucial component of the campaign in obtaining a strong salience on the issue was acts of **civil disobedience.** These acts resulted in news coverage and, in a way, represented dialogue between politicians and ordinary campaigners, who felt that their only other means to communicate with senior officials, namely the public inquiry, was deeply flawed or insufficient. These acts of civil disobedience, which some have labeled direct action, were crucial in not simply sparking greater local opposition but actually forcing the policy position to be considered at the national level. It was able to achieve this, in part, by getting national and international media coverage.

One hypothesis as to why the campaign was able to achieve this level of media attention, which then impacted policymakers' considerations, was that Heathrow Airport was such an important landmark that it naturally attracted publicity. Sarah Clayton of the anti-aviation expansion network AirportWatch stated that:

> A runway at Heathrow is kind of different than to, say, expand at Birmingham or Manchester or something, it's just in a class of its own...[and] because it was Heathrow, the nation's most exciting airport, the media took a huge interest, it always takes a huge interest. You just have to lift a finger and someone takes an interest in Heathrow. It's amazing.... Stories pop up and the press follow it. Same kind of things happening at another airport, the press [do not follow it] (Interview with Sarah Clayton, 18 June 2012).

However, Rootes (2012) noted that despite the airport's greater national salience, until the third runway campaign it had always remained a local issue, with mobilization and policy consideration largely occurring only at the local level. Indeed, John McDonnell noted that he had been involved in campaigning against expansion at Heathrow since at least the 1980s and had not seen anything like the media attention around the third runway campaign (Interview with John McDonnell, 21 June 2012).

McDonnell insists that 'what tipped the scale was the climate change issue.... It spun from protecting our local community, protecting London and...and West London particularly, into...protecting the globe against climate change. So, it snowballed that momentum' and moved the campaign from something that could be labelled as **NIMBY**, or 'not in my backyard', to an issue of global concern (Interview with John McDonnell, 21 June 2012).

This is supported by Rootes' explanation of campaigns that start locally and expand to national campaigns. 'One major factor influencing whether or not the issues of local campaigns are translated into national mobilizations or become national issues is whether or not those issues are already inscribed in salient public policy or are numbered among the campaign priorities of national movement organizations/NGOs' (Rootes, 2013, 106). John Stewart argued that climate change had national salience with the public, with larger national environmental organizations, and with policymakers, but it also attracted the 'urgency and vibrancy' of climate change activists who engaged in civil disobedience (Stewart, 2010, 10). It was the confluence of these media-attracting actions and the nationally salient issue that led to widespread national and international media attention. Groups such as Plane Stupid and Climate Camp campaigned in opposition to the third runway because of climate change (see Chapter 7).

Climate Camp itself received wide coverage during their protest camp at Heathrow.

> [Media attention of Climate Camp] was unprecedented before or since.... Every morning you had a satellite truck and...correspondents from every major newspaper, ...we were constantly doing interviews all day long for ten days straight and from all around the world. So it not only attracted British press attention, but quite major news outlets from all around the world were also interested, which was really surprising to us (Interview with Hannah Garcia, Climate Camp activist, 16 July 2012).

The form of action taken by these groups was important in generating significant publicity, partially because it threatened disruption at the country's largest airport.

As we can see in Figure 4.3, the first big spike in media coverage was the climate camp in August 2007. Media attention increased again when the consultation concluded and in reaction to the Terminal 5 flash mob. The largest peak occurred in January 2009, when the government made a decision on the runway and when Climate Rush held their 'sit-in' at Terminal 1. However, following the climate camp, more national ENGOs participated in the campaign, and there were increased discussions occurring at the highest levels of the major political parties. By making it a national issue, campaigners were able to reach *policy consideration* of policymakers at the national level who were in the position to determine the future of the runway. Policy consideration increased even more so following the Conservative Party's opposition to the third runway.

When it came to *political support*, the campaign already had the Liberal Democrats on their side. The campaign was not particularly important in the Liberal Democrats' position on the issue, but was it important for the Conservative Party opposition? Would the Conservatives have opposed the third runway if no campaign existed, particularly a campaign focused on climate change? The counterfactual must include an

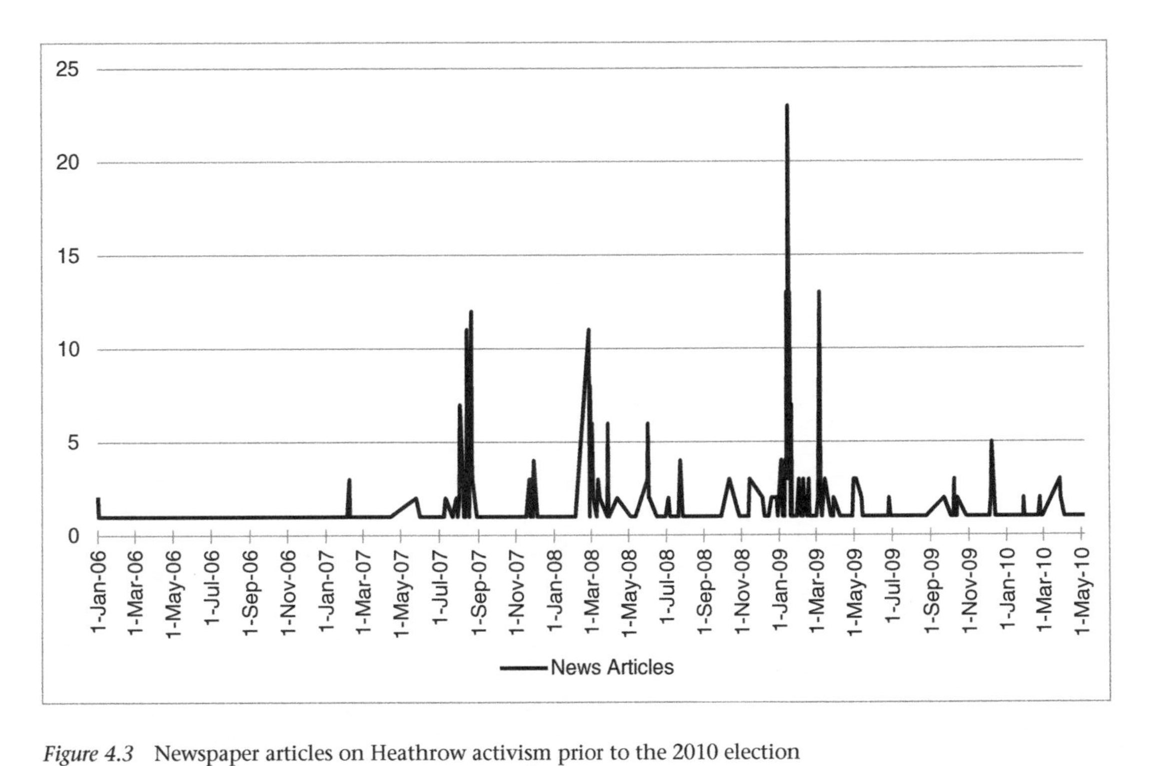

Figure 4.3 Newspaper articles on Heathrow activism prior to the 2010 election

Source: Nexis search: 'Heathrow' and 'third runway' and 'protest' or 'demonstration' or 'activist' between January 2005 and Election Day within UK national newspapers. Duplicates turned off.

examination of the various disruption costs, or the costs placed upon policymakers by campaigners, and concession costs as well as costs that could have been calculable had no campaign been present.

A few months prior to the Conservative Party's announcement opposing a third runway, they had called on the Labour government to support a climate change bill. The Big Ask campaign calling for the bill was national in scope, organization, mobilization, and media attention. A third runway at Heathrow may have been similar to previous Heathrow expansion plans in its relatively localized interest, even though the third runway would have been larger in its consequence to the local community. We can see the lack of attention during the early stages of the campaign. Although the air and noise pollution would have affected two million people in West London, media coverage remained relatively local until the climate issue was addressed. The climate change frame and media attention generated by protests was, according to campaigners, the reason for its national interest (Interview with Hannah Garcia, Climate Camp activist, 16 July 2012; Interview with John McDonnell, 21 June 2012; Interview with John Stewart, 9 December 2011). My personal correspondence with Zac Goldsmith (15 September 2014) also supports this argument. He noted that the climate change frame was crucial to the Conservative Party U-turn on the issue (*political action*), with the expansion impacting 'vast numbers of people, the majority of whom are voters'.

It is reasonable to assume that local concerns were also part of the Conservative Party calculation, considering that the local area had several marginal constituencies in which elections for local MPs were close and could swing from one party to another. Based on data from the Electoral Reform Society, seven nearby constituencies were considered marginal, two of which were close races between all three major political parties.[26] The Conservative Party was able to win five of the seven including both three-way seats (see Table 4.5). However, it must be noted that the Conservatives made significant gains, and Labour had significant losses in a wide variety of constituencies across the country unrelated to local Heathrow expansion concerns. Also, while the constituencies of Carshalton and Wallington, as well as Watford, were under the flight path and affected by noise, they were significantly further away from the airport than the others, and the voters may have not been influenced by a position on Heathrow. While it may be hard to gauge the actual influence of the issue of Heathrow on the electorate in these constituencies, Goldsmith (Personal correspondence, 15 September 2014) noted that local air quality concerns was another reason for the Conservative policy shift, indicating that some calculation was made concerning Conservative candidates in those areas.

Table 4.5 Voting data on marginal constituencies near Heathrow Airport

CONSTITUENCY	2005 Con %	2005 Labour %	2005 Lib Dem %	2010 Con %	2010 Labour %	2010 Lib Dem %
Brentford and Isleworth	30.7	39.0	23.0	37.2	33.6	23.7
Carshalton and Wallington	37.5	17.3	40.7	48.3	8.7	36.8
Ealing Central and Acton	31.2	33.4	30.6	38.0	30.1	27.6
Hammersmith	34.0	42.4	19.0	36.4	43.9	15.9
Harrow East	38.6	45.5	14.2	44.7	37.6	14.3
Richmond Park	39.6	9.2	46.7	49.7	5.0	42.8
Watford	29.6	33.6	31.2	34.9	26.7	32.4

Source: BBC, 2010.

It is also important to note that there were costs to the party in taking the policy position. Namely, it threatened the business credentials of the Conservative Party, and BAA in particular was interested in working with a government that 'wants to listen to business and get business feedback' (Interview with BAA representative, 4 July 2012). Many back-bench Conservatives felt betrayed by the policy move. This calculated cost was not overseen by campaigners. In fact, John Stewart stated: 'I don't think the Conservatives as the traditional party of business would have opposed the third runway…if the economic case [against expansion] wasn't strong' (Interview, 9 December 2011). Therefore, the campaign commissioned the economic report on the third runway that showed expansion was not vital for the economy, a report later used by Cameron in arguing his position on the matter.

Without the economic argument and without major national coverage, that is to say, without the campaign, it is difficult to say if the Conservative Party would have moved on the issue. The party would have received a smaller boost to their green credentials without the national coverage, and they would have suffered more on the economic issue without the campaign's research. While there was still the issue of air and noise pollution, these were also concerns raised by campaigners. Regarding noise pollution, Stewart even argued that the issue was unlikely to garner much political attention:

> I'm not sure that AirportWatch has succeeded yet in getting across to decision-makers and the wider public how debilitating aircraft noise can be for some people. In part this may be because only a minority of noise campaigners have shown the same urgency and vibrancy which has characterised the climate change movement, despite so many local people in the campaign groups being deeply affected by aircraft noise (Stewart, 2010, 10).

If the campaigners were not showing urgency, it was unlikely that many votes from areas unaffected by noise could have been won over to the Conservative Party on those grounds. Climate change provided the framework for making a big 'green' step that resonated with the wider public. Campaigners provided that climate change frame.

The Conservative Party took the first *political action*, changing their party policy to oppose the third runway. While this was a weak form of action, it took place at the highest ranks of the opposition party and fueled a political debate. This led the Labour Party to begin considering its own position on the matter, but the Labour leadership was largely unyielding on a third runway. Despite McDonnell's lobbying efforts, he felt that the party's 'incestuous', 'revolving door' relationship with BAA had 'hooked' them (Interview with John McDonnell, 21 June 2012). In addition, both Tony Blair and Gordon Brown were more interested in market-based job creation than environmental policy.

> Blair never took a detailed interest in any particular policy area. Brown, from when he was Chancellor of the Exchequer, took a more detailed interest in industrial policy. They took the view, which the aviation industry gave to them, basically, that expansion of the airports and expansion of the aviation industry equals expansion of employment and, therefore, was good for the economy, and took that at face value. Neither of them had an interest [n]or understanding of the environmental implications of that, and neither of them ever had a clear view of what the long-term interest of the aviation industry was (Interview with John McDonnell, 21 June 2012).

Nevertheless, some ministers, especially Ed Miliband and Hilary Benn, lobbied the leadership. They put enough pressure on Prime Minister Brown to make significant concessions on the number of flights and aviation standards for the third runway. Ed Miliband, who had been a protégé of Brown's (Darling 2011, 111), had nearly quit over the issue, breaking his relationship with Brown (Seldon and Lodge 2011). The actions taken by Miliband and Benn were perhaps indirectly influenced by the campaign, increasing the salience of the issue as well as drawing the Conservative Party to their side, but perhaps more directly through the lobbying efforts of backbench MPs and campaigners like John McDonnell. The coalition government's formal rejection of the third runway represented the next political action, but this fell in line with existing policy positions of both the Conservative and Liberal Democrat parties at the time so did not represent a campaign outcome. Likewise, the Labour Party later changed their position to oppose the third runway, but that was after the campaign had largely come to a close and after Ed Miliband became Labour Party leader.

The campaign provided enough incentives and reduced the level of disincentives for the Conservative Party to oppose the expansion at Heathrow, which resulted in the coalition agreement opposing the third runway. The ability of the campaign to influence policymakers was acknowledged privately by the Labour government, which created a 'Project Delivery and Risk Report' that was obtained by Greenpeace through a Freedom of Information Act request (Greenpeace 2009). The reports obtained by Greenpeace were DfT's assessments of risks that could derail or postpone the third runway project. The reports covered these risks on roughly a monthly basis for the period of November 2007 to November 2008. They indicated the 'inherent risk', including the impact, likelihood, and exposure of the risk; examined the measures that were in place to manage the risk; and included the 'residual risk'. Each component of the inherent and residual risks was determined on a three-point scale: high, medium, and low.

The risks in the report were diverse and included airspace design concerns, air quality and pollution targets, consultation or report delays, court challenges, loss of third-party support (for which they explicitly indicated the business group London First), and new information from recent analyses. For our purposes, three important risks were considered by the DfT civil servants: 'Strength of opposition from residents under flight path in relation to noise and pollution undermines consultation', 'Direct action by opponents of Heathrow expansion leads to short-term disruption at Heathrow and negative publicity' and 'Gov[ernmen]t loses the economic and CO_2 arguments on [London Heathrow expansion]'. Another relevant risk had just closed in the November 2007 report: 'A number of external activities (such as security, fuel supplies, border control, *climate change camp,* and industrial relations) at Heathrow could occur over the summer. Lack of clarity over areas of responsibility or communication messages leads to possible loss of service, delays or damage to reputation' (emphasis mine). This was marked as having medium-level impact but a high level of likelihood and exposure. They noted, however, that Climate Camp had finished and that no other 'activities' were known to them.

Concern over local opposition was marked as having high impact, high likelihood, and high exposure. High risk across the board only appeared in two other active risks. One was the concern over achieving air quality targets, and the other was completely redacted from the report. The DfT's measures to deal with local opposition were threefold: explaining the impacts of expansion and the consultation process through local exhibitions; attending local meetings and meeting with resident associations; and holding meetings with local policymakers and MPs. These steps did little to address the risk, according to the DfT's own assessment, and all three components of the residual risk remained 'high'. This had remained the same from the previous month, although prior data was unavailable. The direct action risk had a medium level of impact but was high both in likelihood and exposure. Here the DfT's measures were to 'strongly manage' each scenario of direct action activities and have a working group to look over the activities in major airports. In addition, they noted that Climate Camp had finished and 'no major or unexpected issues with regard to consultation occurred but confirmed strength of opposition to airport expansion'. The residual risk was, therefore, downgraded to being of medium likelihood and medium exposure. Both of these risks, drawn up concerning the consultation, were closed in the subsequent December assessment, noting that the consultation had been launched and that new risks were being drafted. By that point, however, the assessment of risk from direct action had increased to highs across the board, with the department's measures only reducing the likelihood of residual risk to medium.

By January 2008, the two risks had been consolidated into one, removing the distinction between local opposition and direct action: 'Strength of opposition to expansion at Heathrow leads to direct action during consultation period'. The risk was assessed as medium across the board with the measures to mitigate the risk (the continuous monitoring of protest, keeping staff informed of protest to minimize disruption, and police presence at roadshow events) reducing the likelihood and exposure of risk to low levels. By February, this risk had gotten 'worse' and had high levels of likelihood and exposure with the department's planned responses bringing this risk down to medium, where it remained until the end of 2008.

The risk of the government losing the argument on the economic and climate change impact of the runway was only introduced into the register in July 2008, and measures had not yet been established to manage it. However, it was judged to be of high impact and high exposure but less likely to occur. Mitigating actions were not identified until December, when 'a clear audit of...evidence (particularly on economic and environmental grounds) to support development' would be updated, circulated, and communicated.

This detailed analysis shows that the government perceived campaigners to be a significant risk. This was also supported in comments by a BAA representative, who confirmed the importance of the campaign when asked if they were the reason for the political parties' decision to oppose the runway, responding: 'Yeah...it's not anything that people don't know' (Interview with BAA representative, 4 July 2012).

It is important to note that the Conservative Party did not change their aviation policy completely and still supported expansion at other airports, but soon after the election the likelihood of a third runway at Heathrow was miniscule. However, there have been increased calls from industry to approve of a third runway at Heathrow and increased interest by the government. It is likely that the next government following

Table 4.6 Policy outcomes model – Heathrow third runway

Policy Consideration	**Important**
Speed of gaining policymakers' attention	**
Length of attention held by policymakers	****
Level of political hierarchy where discussion takes place	***
Salience relative to other concurrent issues	**
Political Support	**Moderately Important**
Number of supporters	**
Strength of support	**
Level in political hierarchy of supporters	****
Support from gatekeepers	*
Political Action	**Important**
Political strength of the action	***
Level in political hierarchy of actors	***
Desired Change	**Moderately Important**
Secureness of policy change	*
Strength of disincentives for not fulfilling policy's mandate	****
Breadth and depth of policy's mandate	**
Strength of implementation	*

* indicate level of importance the campaign had in attaining each sub-component. These are averaged to determine the level of importance of each component.

the 2015 elections will again determine the short-term fate of the third runway, and the next wave of activism, which is already gearing up, may again have to attempt to prevent the airport's expansion.

Green Investment Bank

The idea for the Green Investment Bank was developed by the non-profit E3G and taken up as a campaign by Friends of the Earth staff member Ed Matthew. He established the group Transform UK, which networked with others to form the campaigning wing for the idea. They had the ear of policymakers across parties and quickly got their support. While the Labour and Conservative Parties added the Green Investment Bank to their 2010 manifestos, the Liberal Democrats were 'flirting with [the idea]' but proposed a broader infrastructure bank (Interview with Ed Matthew, 21 September 2012). Nevertheless, the Liberal Democrats had been seen as the greenest of the large parties at the time (Friends of the Earth, 2010a), and campaigners felt that 'would hopefully ensure that it had a green focus' (Interview with Ed Matthew, 21 September 2012).

The Liberal Democrat idea predated the campaign around the Green Investment Bank, but the campaign was still important in providing the *policy consideration* for a *green* investment bank that previously was not considered by the parties.

Quick support did not, however, result in the quick legislation of the policy. The first Queen's Speech following the elections did not promise the GIB, and it was only in the next Queen's Speech in 2012 that the bank was announced. The delay, and subsequent government conflict over the bank, was tangled in concerns over the economic crisis, which placed the bank on a lower priority for the coalition government.

The speed at which the policy was supported by the two largest political parties suggests that the campaign's impact was in the area of policy consideration rather than pressure on *political support*. Once the parties had a chance to consider the policy, they quickly got on board without the need to be lobbied heavily on the issue. While the parties may have been happy with the policy idea immediately, it may have been important to show it had support from the financial sector. This was Aldersgate Group's major role in the campaign, and they were able to attract support from a wide variety of sectors, which the group's organizational structure provided for:

> It's a membership organization. But what's unique about it is its breadth of membership.... It has an environmental focus but very much from the point of view of the wider economy, green growth, the green economy and all of the issues around the transition to a carbon resource efficient sustainable economy. And because of that its objective is to have as much influence as it can, but not to grow as large as it can. So the sort of philosophy...has been to restrict our business membership to really a couple of leading companies in each sector, to get as many sectors on board as possible. That means we need to keep reasonably small. It means that we can have a collegiate approach of consensus-building and finding where the common ground is... (Interview with Peter Young, Aldersgate Group, 8 November 2012).

By getting the business community on board, the campaign signaled to policymakers that the bank was an economically acceptable institution. This may have indirectly spurred policymakers' support. In addition, while Aldersgate Group's alliance of industry representatives did include a few members that were negatively affected economically by the GIB, they had 'never been directly interfacing' with the campaign (Interview with Peter Young, Aldersgate Group, 8 November 2012).

Campaigners felt they had support from Liberal Democrats deputy prime minister Nick Clegg, business secretary Vince Cable, and energy and climate change secretary Chris Huhne and worked to 'build relationships with all those different departments and the ministers' (Interview with Ed Matthew, 21 September 2012). 'Nick Clegg's office actually was very keen to talk with us about the development. Nick Clegg seemed very keen that the institution should survive and wanted to know what needs to be done to take it forward' (Interview with Ed Matthew, 21 September 2012). The campaign had frequent communication with Cable's advisor, and Huhne was seen as being 'absolutely critical in getting support for us for the institution. He played a very strong defensive role as George Osborne was beginning to [go] against this idea of a green investment bank' (Interview with Ed Matthew, 21 September 2012). However, their support was not as strong as the campaign would have liked.

The support of Vince Cable was viewed skeptically due to his desire for a more broad investment bank:

> [The GIB] wasn't a particularly big priority for him.... [I]t turned out that he was actually prepared to trade away the Green Investment Bank to get purchase from the Treasury department policies, so he was perfectly prepared to kill it off when it came down to negotiations with the Treasury (Interview with Ed Matthew, 21 September 2012).

Chris Huhne's support had wavered when pressed by the Treasury. After cryptically condemning the Treasury's position on borrowing, he backpedaled, stating,

> Obviously, if we were to turn around and have the GIB borrowing vast amounts of money tomorrow I can understand that managers of the national debt would be a little alarmed by that. I am absolutely at one with the Treasury on the need to make sure our fiscal credibility is completely re-established.... Let there be no doubt that the first overwhelming priority of the government has to be to get the deficit down (as quoted in Stratton and Webb 2010).

In addition to failing to muster significant support from policymakers, the campaign was unable to get any support from the Treasury and from Chancellor George Osborne, the key gatekeeper concerning the powers of the bank. It was not due to a lack of effort on the part of the campaign. They had met with the Treasury and the meeting 'had gone pretty well', but after the Treasury decided to oppose a strong bank, they 'closed down' communications and would not speak with campaigners (Interview with Ed Matthew, 21 September 2012).

Even after campaign lobbying efforts and the eventual agreement by the Liberal Democrats to adopt immediate borrowing powers for the Green Investment Bank as a party policy, the campaign was still unable to effectively overturn the Treasury's position. Concordantly, Liberal Democrat chief secretary to the Treasury Danny Alexander did not appear to press the Treasury on the issue:

> [H]e seemed to be doing almost nothing to defend [the bank]. I don't know why this was because he's been almost impossible to connect to or communicate with during this. It seemed like he'd completely gone native as far as the Treasury political ideology was concerned... (Interview with Ed Matthew, 21 September 2012).

Campaigners felt that the ideological factor was the primary reason that the Treasury would not support a proper Green Investment Bank. They felt that key Conservative ministers were ideologically opposed to a public bank and in favor of cutting public spending (Interview with Ed Matthew, 21 September 2012; Interview with David Powell, Friends of the Earth, 18 October 2012). It was noted that even if they had the ear of the Treasury, the ideology was so 'firmly entrenched' that 'a chat is unlikely to budge them' (Interview with David Holyoake, Client Earth, 25 October 2012).

When it came to *political action*, the Conservative and Labour parties quickly moved on the issue by pledging a green investment bank. While it was evident that the campaign provided the parties the idea for the bank, the actual action was quickly taken without much pressure on the part of the campaign. The Conservative-Liberal Democrat coalition agreement included the Green Investment Bank following calls for the new government to do so prior to the elections. Again the campaign was not able to get strong political action, with the Green Investment Bank reappearing in the Queen's Speech two years later. When the government produced the draft bill, it was weak and the campaign was closed off from lobbying the Treasury.

With the support of Labour MP Iain Wright, Client Earth's amendments were introduced in the House of Commons but none were approved. These included amendments at the committee phase, where votes split down party lines with the Labour committee members voting in favor and Liberal Democrat and Conservative members voting

Table 4.7 Policy outcomes model – Green Investment Bank

Policy Consideration	**Important**
Speed of gaining policymakers' attention	*****
Length of attention held by policymakers	**
Level of political hierarchy where discussion takes place	**
Salience relative to other concurrent issues	**
Political Support	**Of Little Importance**
Number of supporters	*
Strength of support	*
Level in political hierarchy of supporters	***
Support from gatekeepers	*
Political Action	**Of Little Importance**
Political strength of the action	*
Level in political hierarchy of actors	*
Desired Change	**Of Little Importance**
Secureness of policy change	**
Strength of disincentives for not fulfilling policy's mandate	*
Breadth and depth of policy's mandate	*
Strength of implementation	**

* indicate level of importance the campaign had in attaining each sub-component. These are averaged to determine the level of importance of each component.

against. While the amendments themselves were clearly influenced by the campaign, with Wright presenting the amendments as written by Client Earth, they were relatively low-intensity actions that did not result in further gains. Likewise, amendments in the Lords also failed.

While the government proposed its own amendment on green purposes, which was likely a result of campaigning pressure, the amendment did not directly tie the green purposes of the bank to the Climate Change Act and did not include the funding of fledgling technologies. As of September 2014, the Green Investment Bank still does not have borrowing powers and is not promised these powers in legislation. With a weak mandate and weak green purposes, the campaign did not seem to have played much of a role in attaining their *desired change*.

Conclusion

We can see that campaigns played some role in every component of the Policy Outcomes Model (see Table 4.8). This role, however, was not consistent between campaigns or across components. Policy consideration

Table 4.8 Summary of policy outcome model results

	Climate Change Act	Heathrow Third Runway	Green Investment Bank
Policy consideration	Very Important	Important	Important
Political support	Moderately Important	Moderately Important	Of Little Importance
Political action	Very Important	Important	Of Little Importance
Desired change	Important	Moderately Important	Of Little Importance

was the component that was most strongly and consistently influenced by campaigners. This is partially explained by the fact that both the Climate Change Act and the Green Investment Bank were campaigns to advance policy ideas created by campaigning organizations. Therefore, they were critical in influencing policy consideration simply by developing the idea and presenting it to the government officials and opposition policymakers. The other components had more mixed results across the campaigns. The campaign for the Green Investment Bank was certainly the weakest in its ability to influence policymakers.

The reason for the differences between campaigns and across components is the subject of subsequent chapters. These will explore how the political opportunities, strategies, and mechanisms influenced policy outcomes in each of the campaigns.

5
Political Opportunities

Social movements do not operate in a vacuum. Their efforts, strategies, and tactics (see Chapter 6) are not the sole determining factors in a movement's ability to influence policymakers and create policy outcomes. Political contexts, processes, and structures all help shape the abilities of a movement to influence policy. This argument underlies the political process approach to social movement theory.

One important concept of the political process approach is **political opportunity structures** (POS). POS refers to the possibilities and constraints that stable features of a political system provide to social movements' mobilizing efforts and outcomes. POS is political in that it is interested in variables found in the political system in which social movements are making claims; it focuses on opportunities by examining the possibilities and constraints that a political system's features afford to social movements; and the examination of relatively stable institutional variables explains the structural component. POS is interested in looking at variables such as the separation of powers, the strength of the executive branch of government, the type of electoral system, the availability of citizen-initiated referenda, and the length of the electoral cycle (among others) to see if they play a role in determining the level of mobilization and outcomes of a social movement. Often, these variables are grouped together to form indicators of 'open' and 'closed' political opportunity structures (for example, Kolb, 2007; Eisinger, 1973; Kitschelt, 1986; Midttun and Rucht, 1994). By and large, it is hypothesized that governments with open structures are more likely to be influenced by social movements, and, therefore, produce more movement outcomes. Closed structures are more difficult to influence and result in fewer movement outcomes.

Despite its definition, however, POS has been utilized to explain non-political and non-structural variables, leading scholars to heavily critique POS for being indiscriminate, as well as overly structural and potentially tautological (Goodwin and Jasper, 2004). While these arguments were most harshly doled out by proponents of a cultural framework, some of their criticisms were also accepted by those advocating a more constrained and cautious approach to POS and the broader political process model (Rootes 1999). This cautious approach promoted the careful confinement of the concept and the usage of alternative concepts when discussing non-political or non-structural variables rather than 'lumping' the variables together (Koopmans, 1999; also see Meyer, 2004), a problem that has appeared in social movement research (for example, McAdam, 1999).[27] By constraining the concept, scholars reduce the risk of POS 'becoming a sponge that soaks up virtually every aspect of the social movement environment' (Gamson and Meyer, 1999, 275).

Unlike POS, **dynamic political opportunities** refer to the constraints and possibilities afforded to social movements as the result of *unstable, erratic* features of the political system. Dynamic political opportunities concerns variables such as the relative strength of a political party, the party competition around a policy area, the locations of key constituencies within a given electoral cycle, the appearance of an important minor political party, the occurrence of a meaningful political event, the relationship between the government or key policymakers and relevant interest groups, the political ideology of the government or key policymakers relative to their own political party, and so on. These variables change with more ease and frequency than POS and, therefore, present another important component to understanding social movement outcomes. Independent of POS, dynamic political opportunities can help and hinder movement activity in their attempts to achieve policy outcomes.

In order to gain a good grasp of the importance of POS in the outcomes of social movements, analyses should look across political institutions and control for dynamic political opportunities. Usually such an analysis looks at movements across space (that is, different countries). Alternatively, a study of POS can explore campaigns seeking to influence different scales of policymaking institutions, such as regional governments, like US states or cities. For dynamic political opportunities, analysis can take place within a single political body (national, regional, or local government) and automatically control for the structural component. In other words, while it is still necessary to control for other important variables, by comparing campaigns that take place within a single political scale and space you rule out the effects of such things as electoral systems and separation of powers. By examining the dynamic political opportunities of three cases focused on changing policy in the UK at the national level, I am able to demonstrate the influence of unstable and erratic political processes on movement outcomes. Additional methods to control the noise from stable variables would be required if an analysis included cases from varying scales and spaces.

In applying the notion of dynamic political opportunities to understand the influence made by the climate change movement on policies regarding the three cases, it is important to grasp the broader political context at the time. In order to do so, I briefly present a historical sketch of the political and public arenas during the time of the cases.

Climate change policy window

It has been acknowledged by analysts and scholars that the period of time in which the Climate Change Act was introduced, Heathrow's proposed third runway was protested, and the Green Investment Bank was initiated fell within a policy window (Carter and Jacobs, 2013). A policy window is a period of time in which opportunities for the passage of legislation on a given policy area are open. Such policy windows open infrequently for any one policy area but, once opened, policy windows are periods in which 'advocates of proposals' (Kingdon, 1995, 165) are most able to influence policy change.

Policy windows provide opportunities for policy decisions to be made and can be used to understand policymaking beyond movement outcomes. However, policy windows can also be seen as dynamic political opportunities for social movements. During those moments, movements can provide policymakers with solutions to policy problems, pressure policymakers to strengthen or redirect existing proposed solutions, or help extend the time in which the policy window is open by maintaining political and public attention on the issue. Movements can also experience an easier time of reducing concession costs and increasing incentives to adopt their preferred policies. While in the US these windows are generally open only for a short time, partially due to the difficulty of passing legislation, in the UK they can be open for longer periods of time because the government generally holds the majority in the House of Commons and, therefore, has a much easier time getting legislation approved. Nevertheless, policy windows can still be quickly closed in the absence of other factors (Carter and Jacobs, 2013).

Stating that an open policy window is important for movement outcomes is not enough. We are also interested in understanding a) what caused the policy window to be open, b) what role the movement had in keeping the policy window open, c) how much of a role the policy window played in the establishment of policies, and d) how much the policy window offset the outcomes of the movement or provided the movement with a political opportunity that it effectively exploited.

The UK political system

In order to understand how the climate change policy window opened, it is important to have some knowledge of the workings of the UK political system. Like the United States of America, the UK has a first-past-the-post electoral system in which the candidates in an election win if they have the most votes without needing to have a majority. However, the head of the government in the UK is not directly elected. Instead, both the lower legislative and executive branches are formed in the general election by voting for local candidates that seek to represent their constituency in the House of Commons. Each constituency elects one MP. While the exact number of constituencies changes due to the redrawing of constituency boundaries, the number has stayed above 600 since before universal suffrage. The party that wins the majority of constituencies controls the government, with the party leader taking the position of prime minister. If no party controls a majority of seats, a hung Parliament is resolved through the formation of a coalition of parties or a minority government. Unlike the United States, only one chamber of the bicameral legislature is elected. The other chamber, the House of Lords, is made up of 'peers' who are generally appointed. There is no fixed number of peers, and their role in the legislature is to scrutinize policy.

The contemporary political landscape of the UK features two dominant parties, the center-left Labour Party and the center-right Conservative Party. The third largest political party, the Liberal Democrats, grew out of a merger between the Liberal Party and the Social Democratic Party in 1988 and consists of a socially liberal strand and an economically liberal strand, often with a center-left and environmental bent. The two largest parties have consistently fought each other to gain the majority of seats in Parliament while third parties attempt to obtain a small but substantial number of seats and challenge the major parties as opposition parties while attempting to gain more popularity and votes in subsequent elections. The Liberal Democrats have been particularly successful in this regard.

Party competition and public opinion

Climate change was being addressed prior to the policy window opening, but mostly on the international level. Under Tony Blair, the UK was a major voice in international climate change debates. However, the UK's own emissions reductions were largely being achieved through the shift from coal to gas that was experienced during the 1980s, unrelated to any climate policy. Aside from demanding greater effort to tackle climate change at the international level, the Labour Party's campaign promises, found in their 2001 manifesto, were to obligate electricity companies to deliver 10 percent of their domestic electricity from renewables by 2010 and spend £100 million (~$70 million) on low-carbon technologies. In 2005, they also indicated that greenhouse gas reductions of 60 percent were 'necessary and achievable' by 2050 (with the help of 'clean coal'; Labour Party, 2005), but they did not propose to legislate this target. In 2005, the Conservatives called for the phasing out of hydrofluorocarbons, a potent greenhouse gas, but said nothing about carbon dioxide. They also called for increasing grants to 'significantly reduce the cost of cars with low carbon emissions' but did so alongside a pledge to reduce a vehicle tax (Conservative Party, 2005). The Liberal Democrats had the strongest position on the matter, calling for a carbon tax as well as other market-based climate solutions in addition to advocating for strong international action (Liberal Democrats, 2005).

Competition between the political parties resulted in the climate change policy window opening in 2006. It is important to note, however, that party competition did not occur around environmental issues previously and, therefore, environmental policy windows were sparse. Carter (2006) argues that party competition did not typically occur due to environmental issues having low salience, not proving to be a factor in general election voting, and not being a mainstay of a particular political party that would have made it a partisan issue. Instead, the environmental issue had historically been downplayed by both major parties who 'still compete primarily along left–right lines, so party strategists need only be concerned about the environment if there is an upsurge of public concern and their major opponent chooses to compete on the issue' (Carter, 2006, 760). However, there was no hint of a surge in public opinion on the issue when the policy window opened.

Whereas surveys showed concern for the environment in the UK to be higher than that of economic, social, and political problems in 1989 and was equally ranked to these in 1990, this was no longer the case in the 2000s (Harrison *et al*, 1996, 216). In a 2002 poll, environmental issues including climate change ranked the sixth-largest problem facing Britain today (Ipsos MORI, 2002). Throughout 2005, Ipsos MORI polls asked respondents to spontaneously state the most important issue facing Britain and 'pollution/environment' consistently placed approximately tenth (Ipsos MORI, 2005a, 2005b, 2005c, 2005d, 2005e, 2005f, 2005g,

2005h, 2005i, 2005j). In 2006, YouGov asked Britons what issues they were most concerned about on a daily basis, and 'climate change/global warming' came in fifth (YouGov, 2006), dropping to between sixth and eighth the following year (YouGov, 2007a, 2007b, 2007c, 2007d, 2007e). In addition, one 2007 YouGov poll found that climate change ranked the fifth *least* important issue for the government to focus on (YouGov, 2007d), and in a survey in 2009, respondents placed 'projects to avert climate change' third highest on a list of preferred places for the government to *reduce* spending (YouGov, 2009).

Surveys tended to show terrorism, crime, immigration, the healthcare system, and pensions having higher priority than climate change and the environment over the years. When the financial crisis hit in 2008, even more emphasis was placed on economic concerns compared to the environment (Ipsos MORI, 2009; also see Carter, 2006). The data suggests that, when contrasted with other issues, 'the environment almost disappears from the radar' (Carter, 2006, 759). That being said, the issue of climate change was still viewed as an important issue in need of being addressed. Since 1998, public opinion polls had shown that a sizeable majority saw climate change as an important issue that required government action (Ipsos MORI, 1998, 2004, 2007). However, some pollsters label this a 'are you a heartless bastard' question (UK Polling Report, 2006a), whereby people are not willing to say that they do not care about climate change in fear of being judged for not being caring individuals. Other evidence suggests that this is perhaps too cynical of a position to take regarding public sentiment on the environment. For example, the level of public support for ENGOs is quite high in the UK. Specifically, 'the aggregated numbers of members or financial supporters of environmental NGOs exceed five million, and almost one adult in five claims to be a member of one or more environmental organisation' (Rootes, 2011, 47). While it is not difficult to claim membership and provide financial support, it still suggests that the environment does have some weight in UK public opinion.

While the data on issue salience did not show that climate change would be the likely or electorally demanding reason for party competition, competition did occur nonetheless. This started after the 2005 elections that occurred just following the introduction of the climate change presentation bill. The election results gave the Labour Party just barely a plurality of the vote (35 percent to the Conservatives' 32 percent), but they received a substantial number of seats in Parliament, giving them a small majority (356 out of 646 seats). Labour's share of the votes and seats in Parliament dropped from the previous election. They had paid the price for Blair's decision to back the Bush administration's war in Iraq, seeing their popular support drop following the insurgency. While the Conservatives made moderate gains of more than 30 seats in the 2005 election, they hardly increased their proportion of the popular vote. At the same time, the Liberal Democrats made a 4 percent gain, obtaining 11 additional seats. Following the elections, the Conservative Party leader Michael Howard announced his resignation. Blair, too, had declared that he would not run for a fourth term as prime minister. The next election, scheduled for 2010, would feature two new party leaders.

In the case of the Conservatives, their next leader needed to produce gains in the popular vote, tackling the problem of their 'nasty party' image and their failure to appeal to younger voters (White and Perkins, 2002). David Cameron was up for that challenge. In his speech to the Party Conference, when running as a candidate for party leader, he called for 'a Conservative Party that has the courage to change' and 'switch a new generation to Conservative ideas' (Cameron, 2005a). Although he was not the favorite to win the leadership vote, his speech made a strong impression, and he was later elected.

The Conservatives were neck and neck with Labour in voting intention polls, and in order to push his party ahead, Cameron stuck to his speech. He and his staff decided that concern for the environment was important to reinvigorating his party. From the outset, Cameron discussed climate change as an important issue (Cameron, 2005b), established an environmental policy group, and met with members of Friends of the Earth and Greenpeace. It was clear that David Cameron was pursuing a 'greening' and rebranding of the Conservative Party. In 2006, with the local elections approaching, Cameron joined WWF on a trip to the Arctic as a publicity campaign to show off the party's new green image. Labour tried to retaliate by producing an ad picturing a chameleon on a bike, Cameron's new mode of transport to increase his green image, with the tag line 'available in any colour (as long as it's blue).' According to polls, the Labour Party's retaliation failed (UK Polling Report, 2006b). A later poll found that 47 percent of respondents felt Cameron genuinely cared about the environment more than other politicians did, but 53 percent did not trust the Conservative Party to implement environmentally friendly policies. More people disagreed than agreed that Cameron's environmental agenda meant the Conservatives Party was 'really changing for the better' (Populus, 2006). These numbers were not substantially better for 18–34-year-olds, who were the most likely to say that they did not trust the Conservatives when it came to the environment. Despite mixed feelings in the polls, the environmental policy window was opened.

Policy solutions from the movement

The opening of a policy window creates an opportunity for policy solutions to be proposed. This allowed campaigners to present their preferred policy positions or frame specific campaigns in climate change terms in order to take advantage of the window. From the angle of political parties, having begun competing on the environment and climate change in particular without a clear policy solution meant that they did not have strong control over the policies they would eventually compete over. Movements could propose policy solutions, and if they were presented by trusted organizations or were salient enough in the public they would become a source of competition. This particular form of policy window is highly beneficial to social movements because not only does it allow room for maneuver, but it also allows movements to *set* the policy solutions during the window.

The policy solutions of the Climate Change Act and the Green Investment Bank were creations straight from social movement organizations. In part, the organizations that were credited with designing or promoting the proposed policies were already seen as trusted members of the environmental policy community. Friends of the Earth 'had a track record in getting new bills through Parliament' (Canzi of FoE, in Hall and Taplin, 2007), and Transform UK was initiated by a FoE staff member and included other organizations with elite contacts. In the case of the third runway, the campaign reframed the issue to focus on climate change and was able to generate enough public concern that it became a political battleground.

Policy window, micro-political opportunities, and the Climate Change Act

FoE had already noted 'a convergence of circumstances' that provided opportunities for running a climate change campaign before any policy window was opened (Interview with Tony Juniper, 18 September 2014). Increasingly, more NGOs were becoming interested in climate change, and the science was gaining more and more ground. Even more importantly, a speech by Tony Blair had signaled his intention to be an international leader on the issue at the 2004 Labour Party conference (Interview with Tony Juniper, 18 September 2014). Following discussions within FoE about what climate change campaign to run, they decided to call for targets at the national level and established the Big Ask campaign (Interview with Tony Juniper, 18 September 2014).

When the policy window opened, the Conservatives were ahead of Labour in the polls, while the Liberal Democrats continued to hold approximately 20 percent of the public's support. In the local elections, Labour had experienced a sizeable defeat, losing 17 of the 176 total councils, with the Conservatives gaining 11 councils under Cameron's leadership. Despite seeing no immediate benefits in the polls, Cameron continued to advance the green cause and, in September, made two moves that further opened the policy window. One was to replace the Conservative Party logo from a torch to a green tree. Prior to that, however, Cameron called on the Labour Party to take up a climate change bill, showing support for a major climate change policy. While this action was weak and allowed the Conservative leadership to look environmentally friendly without having to pass the legislation themselves and deal with party rebels, it resulted in increased party competition around climate change. Party rebels and traditional party members were enough of a constraint on Cameron that his announcement to support such a broad-ranging policy was made alongside Friends of the Earth – both shielding Cameron from criticism and helping to promote the issue amongst those who trusted FoE.

From the movement's side, FoE was aware that the policy window was expanding (Worthington, 2011), but they did not restructure their Big Ask campaign in light of this information. While the Conservatives were clearly looking for a policy to stand behind, campaigners were hesitant to rely on any one political party to take up their climate change bill, knowing that such windows often close quickly (Kingdon, 1995). Additionally, had FoE tried to persuade only one party to take up the bill, the party may have agreed for electoral expediency but later U-turned on the issue, or simply failed to be elected into government. Instead, FoE worked to get a cross-party consensus on the issue, and at roughly the same time (see Carter, 2006). If every party agreed to the policy, it would have been a policy promise no matter who formed the next government and would have less trouble going through Parliament (Interview with Tony Juniper, 18 September 2014). In addition, creating party competition around a policy at a time of climate change policy window was an attempt to keep that window open to ensure the act was passed.

Cameron's support for the campaign provided additional pressure on the Labour government to keep their green edge over the Conservatives as well as the Liberal Democrats who were also attracting Labour voters (Johnston *et al*, 2006). Bryony Worthington noted that Environmental Secretary David Miliband 'was quite skeptical about needing legislation, but there was David Cameron saying he would deliver a bill, so very quickly it became government policy that they would also deliver a bill' (Worthington, 2011).

With Blair announcing that he would resign in the near future, David Miliband had perceived that his position was under threat in the next cabinet reshuffle. According to Bryony Worthington, this made him call for the bill to be quickly drafted, presenting a **micro-political opportunity**. Due to the speed in drafting the bill, it encountered less scrutiny by its two major opponents, the Treasury and Department for Business, Enterprise and Regulatory Reform.

> We ended up arguing with the Treasury more on the principle than on the detail because we were moving so fast that they only had maybe one or two policy people covering our brief whereas we had, you know, a team of lawyers and us and all our special advisors and we basically just were able to outwit them a little bit by moving quickly (Worthington, 2011).[28]

A micro-political opportunity is an opportunity that, rather than presenting a broader political shift, creates a small opening that is exploitable by the movement and has some effect, in this case concerning policy outcomes. The pressure felt on a policymaker's position allowed a movement insider to advance a movement-favored policy.

Despite the micro-political opportunity and larger policy window, the government's climate change bill was not satisfactory to campaigners, who continued to pressure the government to adopt measures to strengthen the policy. While the continuation of the open policy window and party competition provided additional opportunities for campaigning, no micro-opportunities were crucial in the strengthening of the Climate Change Act. While the Conservative Party advocated for a climate change act, they did not show active support for strengthening the Labour government's bill except on the position of annual targets (Ares, 2008). Liberal Democrats, on the other hand, tabled an amendment for the 80 percent target, but it was handily defeated 53 to 150 (Ares, 2008). It was the Committee on Climate Change, in addition to pressure by NGOs, which led to the adoption of an increased reduction target.

This section has shown just how important political opportunities were for the Climate Change Act. This was reiterated by FoE executive Tony Juniper, who said that the Big Ask campaign 'underlined the importance of context...because you could have run that campaign three years before or three years after and it would not have worked' (Interview, 18 September 2014).

Policy window duration and cumulative outcomes

Even after the passage of the Climate Change Act, party competition on the environment was sustained. When Blair resigned in late June 2007, the Labour Party experienced a boost in the polls, pulling votes away from the Conservatives. Blair's Chancellor of the Exchequer, Gordon Brown, became party leader and prime minister, with his only competition, left-wing MP John McDonnell, unable to get enough nominations to run against Brown in a leadership election. Despite Labour lining up behind Brown, Blair left behind a divided party on several issues, and despite the boost, the Labour Party quickly lost its lead. The Conservatives had again achieved a plurality of popular support, which increased steadily until late 2009 when the Conservative lead began to decline. Polls in the few months before the elections in May 2010 predicted a tight race, with the Conservatives having only a slight lead over Labour and the Liberal Democrats, who were polling neck and neck.

In the 2010 elections, the Conservatives increased their popular vote by 3.7 percent and gained 97 seats. Labour lost 91 seats and 6.2 percent of the popular vote. Although the Liberal Democrats increased their popular vote by 1 percent from the previous general election, they lost five seats. The election resulted in a hung Parliament as no party had a majority of seats – only the second hung Parliament in the UK since World War II. The Conservatives, getting the greatest number of seats, began to talk with the Liberal Democrats on forming a coalition. Prime Minister Brown resigned, and Cameron was invited to form a coalition government, which he did with the Liberal Democrats.

Sustained party competition around climate change resulted in a surprisingly long-lasting policy window that stretched across policy solutions. In part, this could have been the effect of such a large policy problem as climate change. Another reason for the long duration of the policy window was the movement's ability to manage and sustain **cumulative outcomes**, or using previous movement outcomes in the service of additional gains in subsequent campaigns. The timeline of the three campaigns are spread out, allowing us to investigate potential cumulative outcomes. The Climate Change Act became law in 2008, while the third runway was abandoned in 2010 following the elections that brought the coalition government to power. The Green Investment Bank was formed in 2012.

The Climate Change Act represented a comprehensive climate change policy and policy solution. It had the capacity to close the policy window once it was passed. Two things prevented this from occurring. First, the Conservative Party was competing on environmental grounds but gained little by pressuring the Labour government to adopt the Climate Change Act. They wanted to continue to make gains using the environment and sought other policy solutions to do so. Heathrow expansion became one such policy. Second, the comprehensiveness of the Climate Change Act provided campaigners with a means to further develop policy solutions to achieve the emissions reductions that were legislated in the Act.

Anti-expansion campaigners used the Climate Change Act to regularly expose the contradiction between the legislated targets and the increase in emissions that would result from Heathrow expansion (Interview with John McDonnell, 21 June 2012). This contradiction was used to shame the Labour Party, which both passed the Climate Change Act and supported airport expansion. For example, following Labour's approval of the expansion while conceding some ground to the 'Milibenn' wing of the cabinet, WWF-UK lambasted Labour saying, 'Heathrow's expansion was the first big test of the government's environmental credibility since the Climate Change Act became law last year. It has failed spectacularly, and by choosing to support a third runway, the government has torpedoed its own flagship policy' (in Strucke, 2009). Research and court cases regarding Heathrow also questioned the policy in relation to the targets adopted in the Climate Change Act, providing fertile ground for criticism. This was even seen as a source of continued party competition, with Conservatives also shaming Labour for their pro-expansion stance (for example, House of Commons Hansard Parliamentary Debates, 2008a, c.63).

The previous success of the Climate Change Act spurred additional success in the campaign against the third runway. This was even acknowledged by policymakers, who stated that the Climate Change Act was a central reason for the failure of the third runway expansion (House of Lords, 2010, c.1286). However, such cumulative outcomes depended on the existence of a policy window. In the case of the Green Investment Bank, which occurred both during the policy window and following its closure, campaigners attempted to tie the aims and objectives of the bank with those of the Climate Change Act. According to David Holyoake (Interview, Client Earth, 25 October 2012), the campaign's proposed amendments were written to balance the environmental harms the bank would mitigate, and the reason for attempting to link the bank to the Climate Change Act, rather than through other means, was the political capital that the act provided. The act was politically expedient (Interview with David Holyoake, Client Earth, 25 October 2012), and it was popular and internationally recognized (Interview with Peter Young, 8 November 2012). Campaigners saw the act as 'the most effective way of making sure [the bank] really does green things' (Interview with Peter Young, 8 November 2012). Existing legislation became a precedent used to strengthen other legislation. This objective failed, as the timing of the GIB campaign coincided with the closure of the climate change policy window.

Framing

While the policy window was open, social movements were able to frame campaigns in ways that aligned them to available political opportunities. **Framing** is the process of engaging with the construction of meaning, and collective action frames are 'action-oriented sets of beliefs and meanings that inspire and legitimate the activities and campaigns of a social movement organization' (Benford and Snow, 2000, 614). Framing acknowledges that social movement actors are engaged in the production of meaning (Benford and Snow, 2000, 613) and that movement organizations attempt to frame the relevant issue in ways that they perceive will resonate with the public and/or policymakers in order to increase mobilization and issue salience. In this way, frames have the potential to influence movement outcomes. Kingdon (1995, 173) described such processes using the example of mass transit:

> When a federal program for mass transit was first proposed, it was sold primarily as a straight-forward traffic management tool. If we could get people out of their private automobiles, we would move them about more efficiently, and relieve traffic congestion in the cities, making them more habitable. When the traffic and congestion issues played themselves...advocates of mass transit looked for the next prominent problem to which to attach their solution. Along came the environmental movement. Since pollution was on everybody's minds, a prominent part of the solution could be mass transit: Get people out of their cars and pollution will be reduced.

Framing was crucial in the campaign against a third runway at Heathrow, with climate change activists promoting the climate change frame among a number of environment and community frames used by the campaign. While the climate change frame existed with regards to airport expansion during previous campaigns and was discussed in scientific research, the connection was not widely associated in the public and in the political sphere. Previously, the environmental concerns around aviation expansion were localized and centered on air and noise pollution. The latter, according to campaigners, was never successfully integrated into public knowledge. As the head of HACAN ClearSkies, part of John Stewart's campaign strategy was to enlist the support of direct action activists concerned with climate change to address aviation expansion's role in increasing greenhouse gas emissions (see Chapter 6). Their actions later increased media attention on the issue, and, with the help of their slogans and press releases, the news media began to couple the issue of airport expansion with climate change.

Whereas third runway campaigners introduced and increased the salience of the climate change frame, campaigners for the Climate Change Act and GIB worked to add additional frames to policies that had prominent climate change frames. The Big Ask campaign took advantage of a political opportunity that was presented by Tony Blair's outspoken position on international climate change policy. Blair was anxious to be a world leader on climate change and initiated key conversations on the topic as host of the G8 summit and when the UK held the Presidency of the Council of the European Union in 2005. In framing the Climate Change Bill, campaigners coupled Blair's motivations with their policy solution, emphasizing the UK's existing role and future leadership on climate change mitigation.

The international leadership frame was clearly a draw for policymakers once the legislation was adopted by the government. While both Tony Blair and Environmental Secretary David Miliband made statements concerning the frame (Friends of the Earth, 2006m), the best indication that this frame resonated with policymakers can be found in the government's draft climate change bill. A quantitative content analysis of the draft bill found 33 mentions of the leadership frame as a justification for the bill. The document itself spans 62 pages, which is an average of more than one use of the international leadership frame on every other page. The international leadership frame came out of a dynamic political opportunity found in Blair's own stated desire for leadership on climate change. The campaign used the opportunity to create 'narrative fidelity' between the government's desired image and the campaign's preferred policy (Benford and Snow, 2000, 622).

Throughout the campaigns, the climate change movement utilized similar, albeit smaller, opportunities often to shame policymakers for contradicting themselves and to frame the campaign around consistency. Once a statement of support was made, a policy was included in a manifesto, or a law was passed, campaigners could use them as opportunities to shame political parties and policymakers into maintaining consistency. This was most apparent in the Green Investment Bank. To maximize political pressure, the GIB campaign often cited previous commitments to a green bank. Campaigners tried to shame the government by pointing out its promises for a strong GIB ('If however the government puts the money in a fund with no ability to raise money on the financial markets it will be too small to make an impact. This would be reneging on the prime minister's commitment to set up a bank and the coalition agreement,' Green Alliance, 2010), as well as for their claim of being the 'greenest government ever' ('So if this government wants to live up to its own billing as the greenest ever, this bank must be independent and properly financed,' Greenpeace, 2010b; also see Friends of the Earth, 2011a).

When Chancellor George Osborne, a key gatekeeper, became the face of the opposition against the GIB, Greenpeace published 'George Osborne's Top 10 Green Quotes' as another means of using frames to shame the government into action, showing that he had promised progress on green issues, including several statements about financing the green economy and supporting the GIB (Greenpeace, 2010c). By framing the bank on the grounds of economic growth and security, campaigning groups were able to play off the chancellor's own remarks:

> As the chancellor has previously said: "I see in this green recovery not just the fight against climate change, but the fight for jobs, the fight for new industry." He passed the first test by increasing spending on low carbon outcomes. To pass the second he has to make the bank one that can lever capital markets and transform the UK's low-carbon infrastructure (Spencer, 2010).

Closure of opportunities

Following the 2010 election, the Liberal Democrats' popularity began to plateau around the 15 percent mark. This was partially due to the Liberal Democrats' support for tuition fee reforms at UK universities, which increased costs for attending university despite a manifesto promise to 'scrap unfair university tuition fees' (Liberal Democrats, 2010). In the meantime, the Conservatives, struggling to deal with the economic crisis, experienced a dip in public support, while Labour, under the new leadership of Ed Miliband, increased their voting intention results, and the two parties polled closely together until Labour took a steady lead in the middle of 2012. At this time, the coalition government was able to adopt ambitious targets to cut carbon by 2027 on the recommendation of the CCC and established the Green Deal to incentivize household energy efficiency. However, by 2013, antagonistic Conservative MPs and concerns around other issues, particularly the economy, led to the closure of the climate policy window (Carter and Jacobs, 2013).

This shift in policy windows can be seen in the framing of the Green Investment Bank. While the campaign for the GIB had always planned to focus on the economic qualities of the bank rather than relying on environmental arguments, the closure of opportunities due to the economic crisis forced the campaign to frame the bank in terms of accelerating economic recovery, coupling it with a new kind of green economy. For example, Green Alliance exclaimed that 'a green investment bank is the single most important step the chancellor could take to secure a low carbon economic recovery' (Green Alliance, 2009), while Friends of the Earth stated,

> [the GIB] could and should be a powerful, throbbing beast of a thing, free to kickstart not just the economic recovery, but a low-carbon, more sustainable recovery at that. It's exactly what Germany has had for years in the KfW bank, which has been one of the major factors in bolstering the German economy in these tricky times (Powell, 2012).

In order to show the prominence of this joint frame, I performed a content analysis of Transform UK's press releases and statements on the FoE (England, Wales and Northern Ireland) website about the GIB throughout the campaign.[29] I coded each of the press releases and web pages for the arguments made in support of the bank and in their appeals to strengthen the GIB, with most sources mentioning more than one of the following 11 frames:

1. Developing a 'low-carbon economy' ('These plans mark a mile-stone in the development of an institution that can play a transformational role in forging a low-carbon economy,' Transform UK, 2011c).
2. Economic growth ('[the] bank could re-power the economy,' Transform UK, 2010).
3. Creating jobs ('A flourishing green investment bank is vital to... create thousands of new jobs,' Friends of the Earth, 2012).
4. Supporting green companies ('Cllr Harvey...called the Green Investment Bank "good news" for green business,' Friends of the Earth, 2011b).
5. Leadership in green technology and energy ('GIB fulfills its potential to help make the UK a world leader in the supply and deployment of low-carbon technologies,' Transform UK, 2010).
6. Fulfilling pledges or existing policies ('The Green Investment Bank is the glue that will hold together the Green Deal,' Transform UK, 2011d).

7 Updating the UK energy infrastructure ('A Green Investment Bank is urgently needed, which should focus its attention on funding an overhaul of the UK's energy networks – the pipes and wires that enable our energy system to function,' Friends of the Earth, 2010b).
8 Increasing investments in renewables ('...create a Green Investment Bank to support renewable energy,' Friends of the Earth, 2010c).
9 Mitigating climate change ('...investment in tackling climate change...,' Transform UK, 2011a).
10 Gaining energy security, independence, and low-cost energy ('It's our long-term dependency on fossil fuels that needs urgent treatment,' Friends of the Earth, 2011c).
11 Opposing dirty energy ('this decision on borrowing drives a dirty, high-carbon truck through the heart of their growth strategy,' Transform UK, 2011b).

Here is a breakdown of how often each of the frames was used in advocating for and defending the GIB:

We can see that the most frequent frames were connected with economic goals. The 'low-carbon economy' frame had the highest number of occurrences and emphasized sustainable development, with economic growth being part and parcel of a plan to mitigate climate change. Following this, the most frequent frame was economic growth itself, arguing that the bank would actually help economic recovery during the recession. The campaigners also argued that the bank would help support green businesses and create jobs, respectively the third and fourth most frequently used frames, tying their arguments again to economic factors.

Table 5.1 Frequency of frames used by the GIB campaign

Frame	**Frequency**
Low-carbon economy	50
Economic growth	35
Support green companies	19
Creating jobs	17
Fulfilling policies and pledges	15
Green technology	13
Update UK energy infrastructure	12
Increase investment in renewables	12
Climate change	11
Energy security/independence/costs	9
Opposing dirty energy	2

As the economic problem began to take priority over climate change in the political sphere, so did the framing of the GIB. Over the course of the campaign, economic frames that were devoid of environmental arguments became particularly important. Looking at the frames that were solely concerned with the economy (economic growth and job creation), we see an increase in their use over time as a percentage of the frames used for the campaign (see Figure 5.1).

Proponents of the political process model argue that political opportunities allow space for movements to maneuver more effectively. When opportunities close, this space also closes, and outcomes become more difficult to achieve. The climate change policy window closed during the GIB campaign, and so they attempted to couple their proposed policy solution to a new political problem (Kingdon, 1995, 172–5). However, the bank's solution did not fit the government's ideology concerning the problem, which was one of austerity tied to the desire to reduce the deficit in state spending. The proposed bank would increase public debt, at least on paper, and, therefore, would not be a politically palatable policy solution.

Campaigners felt that while the GIB was mandated in the coalition agreement, gatekeepers stopped it from being established as a bank on ideological grounds, driving the government's economic policy:

> [T]he decision to stop the bank from borrowing is linked to their desire to control public finances as the responsible thing to do whereas in fact the reason they are stopping the bank from borrowing is because they are ideologically opposed to the creation of public banks and the bank can in fact play a critical part in generating growth in the UK (Interview with Ed Matthew, 21 September 2012).

The Treasury saw the bank as just another program that 'has to be subjected to the austerity drive....' (Interview with David Powell, 18 October 2012). When this ideology was established within the Treasury, opportunities closed and campaigners' discussions with the Treasury ceased. Even policymakers that supported the GIB began toeing the Treasury line (Interview with Ed Matthew, 21 September 2012).

While the GIB was eventually established, it lacked some of its crucial features, particularly those which clashed with Treasury ideology. Although campaigners were hopeful that the ideology was 'coming under extreme pressure' and more powers would be granted to the bank (Interview with Ed Matthew, 21 September 2012), enough pressure never materialized. With the policy window closed, cumulative outcomes like tying the bank's green purposes to the Climate Change Act had little government support. The GIB became perceived as a small concern

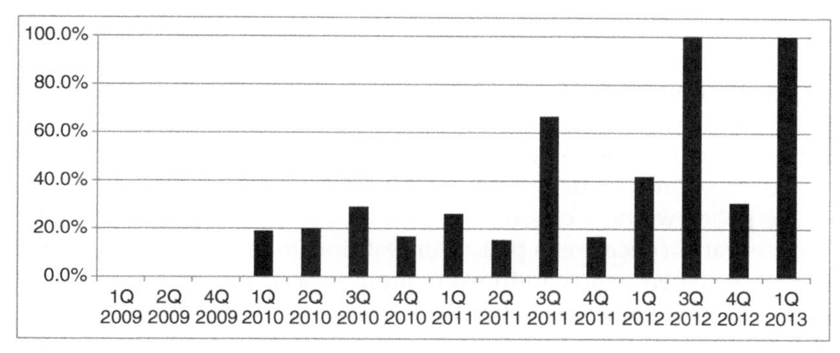

Figure 5.1 Percentage of economic frames in the GIB campaign over time
**The 3rd quarter 2009 and the 2nd quarter 2012 had no data that produced coded frames and were therefore removed from this graph.*

amid more immediate economic problems and, although it was a policy mandated in the coalition agreement, the government created the bank in such a way that it lacked the central features desired by the campaigners, effectively just restructuring existing environmental funds.

Conclusion

It seems clear that dynamic political opportunities, particularly those that opened a policy window, were important for movement outcomes. While the Liberal Democrats were the party to propose more radical climate change policies prior to the policy window opening, none were seriously considered. It was only when the policy window opened in 2006 that similarly radical policies were adopted by the government or called for during the 2010 election cycle.

While climate change was not a highly salient issue relative to others, mitigating climate change still had a high degree of public support, especially for the demographic the Conservative Party sought to mobilize. Environmental organizations reduced the risk for party competition on the issue by holding significant levels of public support and trust. Once the Conservative Party, under the leadership of David Cameron, began to rebrand its image in a green direction, party competition and a policy window followed.

The nature of the policy window, with its focus on climate change but without having a specified policy solution, provided campaigners with the opportunity to create party competition around their preferred policies, granted that the campaigners were able to propose policy solutions that were palatable to the political parties, were salient, or were proposed by ENGOs that had high levels of public trust. After which, a combination of movement and political opportunity factors influenced the strength and durability of the policies. The duration of the policy window was influenced by ongoing party competition as well as through the pressure to maintain policy consistency following the passage of the Climate Change Act, creating cumulative outcomes.

The policy window closed following increased salience of economic concerns and a decrease in policy competition around the environment. GIB campaigners, who continued to promote a strong bank at the time, experienced a change in relations with the Treasury and were ultimately ignored, failing to capture their desired amendments to the government's version of the bank. Campaigners tried to readjust their framing to follow the new policy window, but their desired bank was in direct opposition to the Treasury's ideology.

This clearly demonstrates the importance of policy windows in providing movements with dynamic political opportunities. However, policy windows do not occur in a vacuum either. Social movements can help to precipitate the opening of windows by developing a strong base of support. While it is outside the scope of this book, it is likely that the environmental movement was able to develop the support of young people and position itself as a 'fresh' idea, and in doing so provided a reason for the Conservative Party to have focused on the environment, and climate change in particular, as a point of party competition.

Even when a movement can be instrumental in opening a policy window, the opportunities that opening generates do not automatically result in outcomes. Movements are still required to devise strategies, apply pressure on policymakers, frame arguments, and mobilize support in order to advance their goals. This is particularly the case when a policy window does not come with a policy solution, allowing for various campaigners to vie for their preferred policy.

In the cases presented, even when the government adopted the movement's policy, its version was weaker than what campaigners proposed, and again the campaign had to apply pressure to strengthen the policy. However, it can be justifiably argued that without the opening of the policy window, the policies discussed here would not have been adopted and passed in legislation even with the same level of mobilization. This has been acknowledged by key campaigners in each of the campaigns. In the following chapter, I examine the role strategies and leadership played in achieving policy outcomes, showing that while opportunities play a role, the recognition of and response to those opportunities are also important for movement outcomes.

6
Strategy, Leadership, and Outcomes

Gamson's classic text on social movement outcomes, *The Strategy of Social Protest* (1975), examines strategic variables, among others, finding that several strategic choices made by movement organizations affected outcomes. His research showed that groups that provided incentives for their members corresponded to achieving change for the group's constituency at a greater rate than those that did not have incentives. Likewise, groups that behaved in an unruly manner, either with violence or non-violent constraints such as strikes and boycotts, were more likely to be successful, particularly under certain circumstances.

Although Gamson's analysis features some serious methodological problems (Webb *et al*, 1983), his investigation of movement variables highlights the need to consider the agency of social movement organizations and the role that plays in outcomes. In other words, social movements and the actors that compose them can decide to behave in a variety of ways that can affect policy outcomes. For example, social movement actors can be more or less aggressive in their actions or discourse; they can focus on **inside tracks** or **outside tracks**; they can seek mainstream media attention, communicate through alternative channels, or seek relative anonymity. Social movement actors face a wide variety of choices in campaigning to influence policies. These choices are part of a campaign strategy.

Strategy 'is the conceptual link we make between the targeting, timing, and tactics with which we mobilize and deploy resources and the outcomes we hope to achieve' (Ganz, 2004, 181). This conceptual link allows movement actors to make choices that are strategically implemented (that is, implemented for the purpose of achieving goals; Maney *et al*, 2012). This follows the analytical approach of game theory in which actors face dilemmas and make choices from the options and knowledge they have in pursuit of their self-interest (Jasper, 2004). Not all movement actors have a well-articulated or highly developed strategy. Sometimes strategic decisions are made implicitly by the structure or ideology of a movement, organization, or group. In other words, some decisions are determined by prior choices (Meyer and Staggenborg, 2008, 209–10).

Unlike structural theories that explain outcomes by arguing 'one version or another of "the time for change was right"' (Ganz, 2004, 178), strategic decisions are choices made by social movement actors and, therefore, represent the role of agency in this analysis of movement outcomes. This means that strategic choices are not made purely by rational instrumentalism (Downey and Rohlinger, 2008), rather preconceived notions and expectations tied to the identity of the participants and the biographical knowledges and experiences of the decision-makers all play a role (Downey and Rohlinger, 2008; Ganz, 2004; Meyer and Staggenborg, 2008, 209–10).

Advocates of this power of agency postulate that social movement outcomes are primarily a result of strategic decisions by actors. Whereas structuralists claim that, regardless of strategic decisions, outcomes occur due to such variables as political opportunities or resources, proponents of agency highlight the idea that decisions made by campaigners can influence the creation of political opportunities, take advantage

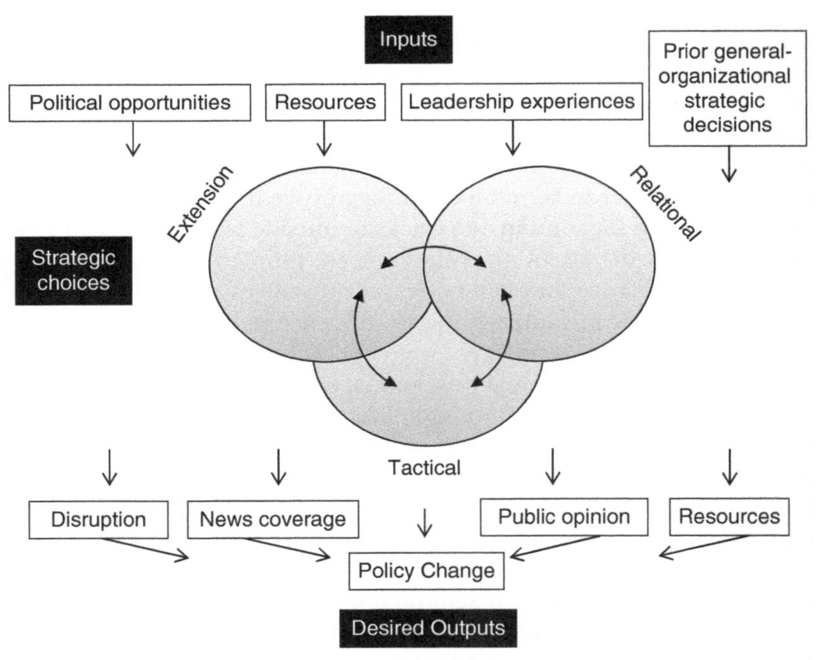

Figure 6.1 Inputs and outputs of strategic decision-making

of policy windows, and effectively utilize resources. For proponents of agency, variables such as resources and political opportunities are viewed as inputs in the decision-making process. Based on these inputs, campaigners can formulate strategies to achieve outcomes (see Figure 6.1).

Despite their differences, however, few structuralists would feel comfortable claiming agency had absolutely no role in social movement outcomes and vice versa. Therefore, we put to the test the relative importance of the agency argument in this chapter, having explored the role of political opportunities in Chapter 5. In order to do so, I will not unpack each strategy of each social movement actor. Instead, I set out to identify key instances of strategic and tactical contributions to the outcomes in the three campaigns.

Strategic domains

In order to accomplish the goal of examining agency through strategy, I have organized strategic choices into categories beyond a distinction of simple (individuals) and complex (groups, organizations, coalitions, etc.) players (Jasper, 2004). While I acknowledge that often 'strategic leadership' will form within a campaign or organization/group that will make the strategic decision (Ganz, 2000; 2004) and that this leadership can be comprised of a single individual, a group of individuals from the same organization, or a group of individuals from different organizations, it is not only important to know *who* made the choices but also *what domains* those choices are seeking to target.

The three domains I distinguish are the general-organizational domain, context-specific-organizational domain, and the campaign domain. The general-organizational domain regards the long-term strategic choices of groups and organizations that will be sustained across numerous campaigns. Decisions regarding this domain may be made during a particular campaign but are strategically assessed regarding the nature of the organization or group. The context-specific-organizational domain is the target of strategic choices that will determine the role of an organization or group in a specific context, often a single campaign, and usually uses the specificities of a particular context to assist in the choices that are made. However, these decisions only pertain to the strategic role of one organization and not the campaign as a whole. Campaign domain decisions are those which concern the whole campaign, across organizations, and are specific to that campaign.

Domains are an important distinction because of the interaction between various strategic choices. As we shall see, one strategic choice is rarely self-contained. Instead, they help determine other strategic choices by limiting options or influencing cost/benefit calculations.

Strategic questions

Movement actors are faced with a wide range of strategic questions they must answer (Jasper, 2004). Here we are primarily concerned with three broad categories: *extension, relational,* and *tactical* questions. Like choices at different domains, the choices made to answer the different strategic questions can go on to influence subsequent decisions. This interaction makes it difficult to discern the influence of each choice on a campaign. Instead I will focus on the overall influence of strategic choices in each campaign.

Extension decisions

Extension questions concern the breadth of organizations or individuals involved in a campaign or group, respectively. The central extension question asks: Should the campaign or group seek to enlarge its constituency, base of support, and membership at the risk of overextending its goals and identity? Most organizations and groups in the three campaigns made strategic decisions regarding the general-organizational domain of extension prior to the campaigns. However, this question still required an answer within the campaign domain.

Climate Change Act

The primary actor in the Big Ask campaign, Friends of the Earth, kept the campaign largely under its control. That is, FoE decided not to extend the campaign very far beyond the organization. This occurred because, in part, FoE felt confident that it could accomplish the task of lobbying for a climate change bill. FoE's decision was made easier through years of effective lobbying efforts and the development of relationships with policymakers (Interview with David Powell, 18 October 2012; Interview with Tony Juniper, 18 September 2014; Hall and Taplin, 2007). FoE was in continuous dialogue with key environmental policymakers, which enabled them to quickly form a cross-party group to call for the legislation.

FoE also had the required financial and human resources to operate the campaign. 'The Climate Change Act campaign was the organizational priority for Friends of the Earth for a couple of years. Probably at some stage half the people at Friends of the Earth were working on it in some way' (Interview with David Powell, 18 October 2012). Tony Juniper stated: 'We put most of our communications team on it, which was pretty big at the time, and most of our local group mobilizers and our parliamentary team, and a lot of our fundraising effort' (Interview, 18 September 2014). In addition, FoE was already structured in such a way that local chapters could engage in lobbying efforts at the local level in a more grassroots fashion. Its high levels of membership and structure made the campaign possible by ensuring that it could generate significant levels of pressure.

Another reason why others were not included in the campaign was that organizations interested in climate change, including FoE, were working in parallel in creating the Stop Climate Chaos coalition, which only launched after the Big Ask campaign had been initiated. While Stop Climate Chaos did campaign to some level as part of the Big Ask, the desire to maintain brand identity prevented a comprehensive, joint effort in pushing an existing FoE campaign (Interview with Tony Juniper, 18 September 2014).

Heathrow Third Runway

The campaign against the third runway was nearly the opposite of the Big Ask campaign. Rather than containing the campaign to the few local organizations that had initiated the opposition, these groups extended participation to a wide variety of groups, organizations, and individuals. It must be said that it is more difficult to contain a campaign in opposition to a policy compared to a campaign that initiates a policy, but campaigners still had a choice as to their own actions concerning extension. In the case of the third runway, these actions were very clear.

John Stewart, chair of HACAN ClearSkies, first helped form a broad and loose coalition of campaigns opposed to airport expansion across the country, called AirportWatch. Later, Stewart attended a conference for veterans of the anti-roads campaign (a campaign that opposed the road-building scheme first initiated by the Thatcher government in 1989 in which Stewart was a movement leader) in order to recruit others. Two activists, Richard George and Joss Garman, were also at the conference:

> [W]hen we went to [the conference] we thought...we might start campaigning around the road-building program. It was really around the edges of the conference, in the breaks and so on, speaking to John...that really made us think aviation should be our focus rather than road building (Interview with Joss Garman, 4 July 2012).

Stewart had discussed with them the climate change implications of airport expansion and 'point[ed] out to us that there were already science reports from people like the Tyndall Centre for Climate Change Research' (Interview with Joss Garman, 4 July 2012). These activists soon formed Plane Stupid and made a strategic decision to join the campaign:

> And just realizing that actually at that point no one was working around aviation, it wasn't in the news, nobody talks about the environmental effects of flying and then we found out about it and we realized that in the UK aviation as a proportion of our country's climate impacts is much higher than other developed countries'.... And so all these kind of conditions...made us think this is gonna be a real front line. We're talking about villages getting destroyed, carbon targets being missed, millions of people already feeling frustrated about noise.... There's huge potential for this to become the kind of iconic battleground that it did eventually become (Interview with Joss Garman, 4 July 2012).

Plane Stupid joined others in a loose coalition of groups opposed to the runway. The formation of a *loose* coalition was a strategic decision made primarily by Stewart, who many campaigners saw as the 'linchpin' of the campaign (Interview with Joss Garman, 4 July 2012; Interview with Hannah Garcia, 16 July 2012). In addition to calling on previous contacts in the anti-roads movement for support, building local alliances, and helping to introduce direct action to the campaign, Stewart was able to capture the imagination of major national organizations (Stewart, 2010, 19).

HACAN, at the heart of the campaign, was a small group with limited resources and, therefore, it could be argued that the 'rational choice' would have been to seek allies and extend the campaign well beyond the small group in order to leverage additional resources. Such an argument could state that rather than being a product of agency, this choice was determined by the structure of the organization. However, previous campaigns to oppose expansion at Heathrow did not pursue such broad extension and did not reach toward allies beyond the local frame. Instead, this was a decision made primarily by Stewart's leadership:

> When we started fighting the third runway campaign, I felt that the group HACAN had to learn lessons from its past defeats because previously..., when I wasn't around,... it fought various developments and not very successfully. We felt the first thing we had to learn was to build up a huge coalition because local residents on their own,

> even if they have got support from one or two local authorities and one or two of the MPs, however strong their argument, in my point of view, are never going to have the kind of power to take on the aviation industry particularly that's backed by government. So that was the reason for forming the coalition (Interview with John Stewart, 9 December 2011).

Indeed, prior research has found that leaders with previous experience are in a better position to form coalitions since a 'leader who is embedded in networks of community and movement groups' already has links to potential coalition partners (Shaffer, 2000, 117). Stewart's experience also provided valuable lessons in making the strategic decision to form such a coalition. If successful, a coalition could provide increased resources; a division of labor; the diffusion of information and awareness, legitimacy, and new frames; and the generation of media attention from different sources (Almeida and Stearns, 1998, 40; Alley *et al*, 1995, 414; Čapek, 1993; Hathaway and Meyer, 1993–1994; Shaffer, 2000; Walsh, 1981). Previous research found that organizations employing moderate tactics or inside-track approaches, like HACAN, were energized by forming coalitions with grassroots groups (Shaffer, 2000, 114). A coalition was important particularly since many local community members, including HACAN members, were hesitant to become more active in the campaign (Stewart, 2010, 24; Interview with John Stewart, 9 December 2011).

Green Investment Bank

The campaign for the GIB was somewhere in between Friends of the Earth's exclusive control of the Big Ask campaign and the broad, loose coalition to oppose the third runway. While the idea of the GIB was the brainchild of E3G and Climate Change Capital, Transform UK and its head, Ed Matthew, were the campaign's surrogate parents. The campaign was held inside Transform UK, which was a hybrid organization and coalition with no consensus on what to call it (Interview with Ed Matthew, 21 September 2012; Interview with David Powell, 18 October 2012). This shows the middle-ground solution to the extension question. While Transform UK was officially an alliance of organizations including FoE, Aldersgate Group, and others, Ed Matthew played a leadership role that gave him a certain level of control over Transform UK:

> [Transform UK is] coordinated by Ed really.... Ed will basically arrange meetings and coordinate what they're discussing...and

> there are meetings of the steering group, which essentially is quite a lot of the...major environmental NGOs, during which we will discuss the politics, Ed will present ideas for what needs to happen, the political interventions that need to happen, and people will commit resources as they're able to do (Interview with David Powell, 18 October 2012).

The types of organizations that were welcomed into the alliance was another strategic extension decision. There was a spectrum of choice that fit the goals of the organization. The campaign was focused on green investments and, therefore, support from the financial and environmental community was needed. An emphasis could have been placed on the support from the financial community or the environmental community or anywhere in the middle. 'We felt that if the idea was being led by green NGOs it was less likely to succeed and that we needed to build an alliance that had business and investors as a core part of it, a central part of it' (Interview with Ed Matthew, 21 September 2012), but, nevertheless, the steering committee included important ENGOs as well as members of the business and financial communities. The need for resources and campaigning expertise likely played a role in the inclusion of ENGOs.

Relational decisions

Relational decisions are concerned with how groups or campaigns see themselves relative to other actors. I distinguish three types of relations: external, target, and intra-campaign relations. *External relations* concern the positioning of the campaign relative to actors outside of the campaign, often the general public or a key demographic or population within a particular geographic space. In making decisions about external relations, groups and campaigns ask the question: Should we try to represent a constituency or change the views of the public? The second type of relation concerns opponents or the target of the campaign. *Target relations* generally concern an individual, group, or set of individuals or groups that hold power, such as government bodies or corporations. Here, groups and campaigns ask the question: Should we have an antagonistic relationship with opponents, or should we seek to influence opponents through more friendly means? In other words, should campaigners attempt to be persuasive or try to pressure their opponents (see Busby, 2010, 39). Finally, there are choices concerning *intra-campaign relations*, in which groups and campaigns ask how accommodating they should be to various groups in the campaign, what kind of hierarchy is formed in the campaign, and what forms of interactions should the intra-group relations have.

Climate Change Act

The Big Ask campaign was fairly clear and single-minded about its approach to external relations. The public was already largely convinced of climate change as a serious concern, and while polls found that it was not a very salient issue, a large proportion favored government action. The public would not need to be convinced of the act's value but would be called upon to help lobby policymakers for support. Specifically, campaign events encouraged individuals to lobby their MPs to support the Climate Change Bill through FoE postcards or online communication (for example, Sheffield Friends of the Earth, no date), calling on MPs to sign the early day motion on the bill or lobby ministers to take action (Friends of the Earth, 2006b). In addition, this strategy attempted to increase membership, and thereby encourage additional action, such as more direct or face-to-face lobbying efforts.

FoE also had a clear approach to target relations, although it had less control over how local branches would engage with those opponents. The Big Ask campaign was heavily geared toward lobbying; campaign members and the public were asked to call on their MPs to take action. FoE engaged celebrities and held concerts to mobilize around the issue but not primarily for the sake of publicity. Rather, the campaign brought people together to engage in various lobbying efforts. Once the bill was taken on by the government, the campaign continued to engage in lobbying for further amendments. This approach took the middle ground regarding the question of pressure or persuasion. Campaigners and public participants used persuasive argument but also indicated electoral pressure when they lobbied policymakers.

While FoE sent various calls to action to their local groups, structurally, those local groups had some independence of action. Nevertheless, even those small units generally engaged in their festive action geared toward publicity and public interest (Friends of the Earth, 2007b) or direct lobbying efforts (Featherstone, 2003). For example, Lewisham FoE held a 'Big Ask public meeting' where community members could listen to local MPs speak about the Climate Change Bill (Burke, 2006). The strategy also worked to pressure and persuade policymakers to make public statements that would reify their support and, therefore, penalize any later U-turn on the policy (Friends of the Earth Birmingham, 2005). This demonstrates that intra-campaign relations were closely aligned. Even though the campaign was largely contained within a single organization, there was structural availability for a diverse set of intra-campaign relations because of the powers of local chapters. However, the tight inter-campaign relationship was more likely a result of a shared sense of identity around the values and goals of Friends of the Earth rather than strategic choice.

Heathrow Third Runway

The campaign to oppose the third runway was much messier than the streamlined and straightforward Big Ask campaign. The relations were as diverse as the organizations and groups that made up the campaign (Nulman, 2015). External relations ranged from connecting the third runway with a wide variety of environmental concerns (including air pollution, noise pollution, and climate change) in the eyes of the wider public, demonstrating to the public the seriousness of these and other local problems that expansion posed, mobilizing and representing local constituencies, and defending the public from additional harmful greenhouse gas emissions. These were decisions made primarily by individual groups within the campaign, with each group often only pursuing one form of external relation.

These choices often reflected pre-existing decisions within the general-organizational domain. For example, HACAN was a local group representing their local constituency, and Climate Camp had been consistent in its position not to target individuals and passengers (Interview with Hannah Garcia, 16 July 2012). For some groups, general-organizational domain decisions were influenced by contextual variables regarding the campaign since the groups themselves grew out of opposition to Heathrow expansion. NoTRAG adopted a local representation strategy while also attempting to increase its membership by working to provide local residents with information about the expansion plans. Plane Stupid initiated its campaign to publicly associate the issue of airport expansion with climate change.

Some groups were conflicted in dealing with questions of external relations. Climate Camp had competing strategies that were attempted simultaneously. One of those strategies was to increase issue salience in the general public using the mainstream media, and another was to communicate through alternative means or through the actions themselves. This strategic choice to engage with mainstream media had been developed over time by Climate Camp activists. Prior to forming Camp for Climate Action, some of the founding members had been active in a protest camp against the 2005 G8 meeting in Scotland. The camp had very little engagement with the mainstream media due to a complete lack of trust.

'[A] lot of people didn't think there should be any level of engagement because they simply paint a stereotype of...protesters..., and we should only be doing our own alternative media' (Interview with Hannah Garcia, 16 July 2012). The media was therefore kept off the camp site. This began to change as their image in the media had not benefited from this strategy. At the first protest camp in 2006, mainstream media 'were allowed on site, once a day for an hour; they had to be escorted and they weren't allowed in all the places and that was less restrictive...[and the] decision to allow that level of access was very controversial' (Interview with Hannah Garcia, 16 July 2012). By the 2007 Heathrow camp, such tensions had eased somewhat, and there was greater acceptance of media in the camp, which arguably increased media attention. This was part of a strategy to engage the general public with their message, increasingly seeing mainstream media as a way of speaking to a mass public (Interview with Hannah Garcia, 16 July 2012).

One campaign domain decision concerning external relations was Stewart's decision to develop the issue of aviation expansion from a local concern to a national one. In doing so, Stewart sought to combat the NIMBY label that other local groups are often given by policymakers and the media (see Rootes, 2007b). Stewart wanted to avoid the label 'because if you have NIMBY factors involved you can't then make the wider arguments' (Interview with John Stewart, 9 December 2011). In order to guard against claims of NIMBYism, Stewart established the AirportWatch network, which served as a communication platform between groups opposing expansion at airports across the UK, and forbid NIMBY groups to become members. This became the network's 'cardinal principle...[:] all of those [groups] work together to say, "no, let's oppose the damn thing everywhere...."' (Interview with Sarah Clayton, AirportWatch, 18 June 2012).

Target relations were also divergent in the campaign. This was partially due to the fact that non-state targets were also involved. Outspoken supporters of a third runway within the aviation industry, including BAA and British Airways, were often targeted alongside the government (Nulman, 2015). Strategic choices had to be made regarding each target, and at some moments those decisions were rethought and new relations were established. Plane Stupid had the most stable relationship, attempting to pressure all three targets consistently throughout the campaign. As the Labour government became increasingly split over the issue following the Conservative Party's outspoken opposition to the third runway, the campaign had to adapt to the changing circumstances. Nevertheless, Plane Stupid protested a speech by Ed Miliband over 'the apparent lack of government resolve for tackling the environmental issues that are crucial for safe-guarding our future' (Plane Stupid, 2009). Others, like John McDonnell, were happy to both appeal to some policymakers and pressure others (Interview with John McDonnell, 21 June 2012).

Climate Camp was consistently questioning its general-organizational and campaign domain decisions concerning target relations. A segment of consistent participants in the camp, largely the autonomists and anarchists, advocated not engaging with policymakers at any level. Instead they would confront corporate and state targets as a means of advancing long-term radical change. Others were in favor of using direct action primarily as a means to force the hand of the government on the issue of the third runway. The disagreements that did arise were not easily solved as Climate Camp's structure did not provide a mechanism for such resolution. They had open meetings and operated in a way that gave everyone a voice. At the same time, it was often the case that so much time was spent on actually planning the protest camps that strategic considerations were rarely discussed (Interview with Hannah Garcia, 16 July 2012).

Intra-campaign relations in the third runway campaign consisted of loose networking. Leadership is important in the maintenance of coalitions (Hojnacki, 1998), and Stewart had played 'a real key convening role between all of the national NGOs and how they worked alongside the grassroots direct action community, and also the local residents' (Interview with Joss Garman, 4 July 2012). Nevertheless, this strategic decision for the loose nature of the coalition largely reflected the decision Stewart made regarding extension.

The relations that formed within that loose network were not always strategic. Division of labor, for example, occurred organically (see Saunders, 2004, esp. 211–2), with some members working to generate media attention (Interview with Tamsin Omond, 17 July 2012), while others produced reports and engaged in more local mobilization. Sometimes groups shared competencies and resources by offering direct action trainings, providing media expertise, and sharing office space (Interview with Hannah Garcia, 16 July 2012). Sometimes groups engaged in strategic partnership with others (Interview with Joss Garman, 4 July 2012; Interview with Tamsin Omond, 17 July 2012). For example, Plane Stupid and Climate Camp both worked to engage with local residents and community members (Interview with Hannah Garcia, 16 July 2012; Interview with Joss Garman, 4 July 2012). Climate Camp enacted an explicit strategy to build rapport with local residents by creating a working group within the camp in order to maintain peaceful coexistence

(Interview with Hannah Garcia, 16 July 2012). Other organizations strategically chose not to form partnerships, fearing co-optation, seeking to maintain group identity or brand image, and disagreeing ideologically (Interview with Hannah Garcia, 16 July 2012; Saunders, 2004, 222–3). While these strategic relational decisions reduced intra-campaign conflict, they did not play an important role in policy change.

Green Investment Bank

Leaders of the Green Investment Bank campaign applied a target relation strategy of persuasion. This was the central strategic decision of the campaign that influenced many other choices. Campaigners thought that the bank 'required the right sort of people saying the right thing to government behind the scenes' (Interview with David Powell, 18 October 2012). What was the campaign's decision as to who qualified as 'the right sort of people'?

> I think it's fair to say, and certainly the feedback we've had from the likes of Vince Cable [support the idea that] it's very, very useful to [have the support of business and financial institutions]...because it brings some of the sectors that intuitively might be reticent about it absolutely on board (Interview with Peter Young, 8 November 2012).

Business and financial institutions, therefore, became crucial to the strategy of the GIB. What were the right things they should say? The campaign wanted the government to be aware that these actors 'think [the GIB is] a good idea. They don't see any problem. They don't see it crowding out any investment from their point of view. They think it's wanted' (Interview with Peter Young, 8 November 2012).

After the GIB was agreed to by the coalition government, the campaigners assumed that the policy would be developed using best practice. They tried to inform this practice by interacting with policymakers as expert stakeholders (Interview with Peter Young, Aldersgate Group, 8 November 2012). This 'capacity building' strategy worked by

> finding out what [policymakers] are aware of and competent of... [because] the Green Investment Bank is trying to do [something] that's relatively unfamiliar so we can rapidly introduce them...[to] people who are going to give them as fast a learning curve as possible because we are very aware of the fact that ultimately the success of the institution is going to be on the quality of investment decisions and that's going to come down on the ability of the officials to set the right project, do the

> right due diligence to understand what the risks are, and the more we can expose them to what we think is the best for the [financial] sector early on, the better decisions we think they're going to make (Interview with Peter Young, Aldersgate Group, 8 November 2012).

When Client Earth was hired by Transform UK, they became part of the coalition, developing its expertise on the issue in an attempt to be more influential in the policymaking process around the GIB:

> [W]e became recognized as an expert stakeholder...and [policymakers] recognized that we were influencing Ed Matthew's strategies and they could see that Ed Matthew's advocacy was working because the amendments [had] become Labour's amendments.... Occasionally they are calling us, bringing us in for help, or our opinion or our advice, to consult with us as the legislation evolves (Interview with David Holyoake, Client Earth, 25 October 2012).

Engaging in a strategy of persuasion explained decisions around external relations. Public mobilization was not seen as strategically valuable and, therefore, the public did not have to be heavily engaged in the issue. Public mobilization was strategically assessed as having relatively little value for several reasons. First, the leadership felt that it would be difficult to muster public mobilization for the campaign. An economic crisis blamed largely on banks made the GIB a hard sell:

> I think it's the reality that we are in the midst of a banking crisis where several national banks have gone belly-up, bankers were disgraced for bringing down the economy. The idea that you would start ... a mass mobilization or activism-based campaign around the development of a bank, it seems anathema to us really. It seems like it was going to be incredibly hard to do (Interview with Ed Matthew, 21 September 2012).

This difficulty was encapsulated in Chancellor George Osborne's joke: '[I]t's the first [time] anyone has ever protested *for* a bank' (Greenpeace, 2010c, emphasis mine).

Second, the campaigners saw the GIB as an important financial institution that would require support from experts in the financial community to show that it was viable. The public was not the target audience that the campaign sought to mobilize (Interview with David Powell, Friends of the Earth, 18 October 2012).

Was such an approach a matter of strategic decision-making or the only viable option given structural considerations, such as limited resources? It was true that GIB campaigners had relatively few resources as compared with the Big Ask campaign, and it certainly takes greater resources to mobilize external actors than to engage in elite-level inside track campaigning if the social capital needed is available. Interviews with members of the coalition suggest, however, that rather than resources determining strategy, it was strategy that determined resources. The campaign first made the strategic assessment that the campaign did not need and would not rely on public mobilization. This led to fewer resources being required and, therefore, fewer resources were called upon (Interview with David Powell, 18 October 2012).

Having 'the right sort of people saying the right thing' also meant intra-group coordination. Like HACAN, Transform UK was at the heart of its wider coalition. However, the GIB coalition was more tight-knit – a strategy of the campaign's leadership, particularly Matthew, who initiated this campaign by assembling supporters. This also meant that there was a greater level of resonance between the coalition partners. A fairly unified message was shared and discussed with the different coalition members, as managed by the campaign leadership, resulting in greater coordination than that which was experienced in the campaign against the third runway:

> That's a deliberate strategy on my part to...speak to the government with one single clear voice rather than end up with lots of stakeholders giving them different thoughts. We're trying to build up consensus and how it needed to develop, and it was my job, if you like, to feed that information through... (Interview with Ed Matthew, 21 September 2012).

Tactical decisions

Tactics are specific means of implementing strategies. If a strategic decision was made to try to persuade policymakers, example tactics could be a) to directly communicate with policymakers, b) to indirectly provide policymakers with particular arguments, c) to expand the arguments being made, or d) to frame arguments in ways that resonate with policymakers. While strategic decisions are broad, tactical decisions are more specific. Tactical decisions also can work like Russian dolls, where one tactical decision (for example, directly communicating with policymakers), leads to further tactical questions (for example, What form of communication? When is the best time to communicate?).

In the three campaigns, a wide variety of tactics were utilized. Even within a single campaign, many tactics can be found. Here I assess the choices of tactics in achieving the campaigns' wider goals, ignoring other tactical decisions that can be made during the course of a campaign. These tactical questions include: What actions will you participate in? How will you plan those actions? How are those actions being framed? Who are the actions being directed toward?

Climate Change Act

In the Big Ask campaign, FoE engaged in a range of tactics, including hosting events, publicity stunts, virtual and physical public mobilization, debates, a video advertisement, and others. Some events were strategically planned by local groups while others were organized at the national level and implemented by the local groups. Such a 'pick-n-mix' approach suggests that only minor strategic thought was given to the particular tactics, with most tactics geared toward similar ends of asking the public to lobby their policymakers.

Perhaps the tactic most strategically implemented was the use of celebrities, from Thom Yorke's consistent outspoken and committed support for the campaign, to a range of celebrities working to create a virtual march. Celebrities were previously used as a tactic by FoE (Juniper, 2008); however, this strategy was particularly used in the Big Ask campaign. The use of celebrities alongside a trusted ENGO brand was seen to generate support, perhaps from those who would have otherwise not been aware or as supportive of the campaign and legislation. There was some sign that the use of celebrities was even successful in directly influencing key policymakers (Juniper, 2008; Event Magazine, 2008).

Heathrow Third Runway

Tactics also varied in the campaign against the third runway. Public consultation was utilized by some organizations, blockades were incorporated in several Climate Camp events, and HACAN lobbied policymakers and decided to commission reports they thought would be tactically useful. Here again, few of the tactics were strategically determined within the campaign. Instead, many of these organizations had made general-organizational domain decisions to engage in particular tactics. Climate Camp is the best example, as the group is structured around a protest camp in an attempt to mitigate climate change. While climate campers may strategize about a wide array of tactics (for example, should we or should we not disrupt Heathrow Airport?), the most important tactic was the camp itself.[30]

The power of Climate Camp, with regard to influencing aviation policy, was its ability to garner media attention. Two variables that influenced the level of media attention were not strategically developed for the Heathrow campaign. The first was Climate Camp's tactical amorphousness, or its ability to make a variety of tactical decisions in a short amount of time. This was a product of the free-flowing and open structure of the organization:

> [The climate camp activists working on media attention and relations] didn't know whether there would be disruption, a) because obviously you can't control the thousands of people who are there, who knows who might go off and do what from the camp as a base or what have you. And b) because the decision of what to do on the day of action wasn't made in advance. The whole idea of the camp was that a group of people would get it up and running and happening and then it was open to whoever's there and the decisions that were made once the camp opened were made by everyone who was there. That included the kind of action that was going to be taken.... (Interview with Hannah Garcia, 16 July 2012).

Media outlets anticipated disruption and began to cover the story. Likewise, the group itself was relatively novel, and both the media and the authorities did not know what to expect:

> The power of novelty...gives protesters a strategic advantage – authorities are unprepared for new strategies, political actors, and themes. Given the inertia of institutional politics, effective responses develop slowly, whereas in the early phase of rapid diffusion, social movements are highly flexible – they appear and disappear in ever-changing guises at unpredictable times and places (Koopmans, 1993, 653).

Climate Camp only held a few camps prior to the Heathrow camp, and authorities were still unsure of their intentions. The same was true of Plane Stupid, as indicated by Stewart in his appraisal of the tactic of direct action:

> It really can bring an additional threat to the authorities because governments, the authorities (particularly the civil servants)...like to know where their...opposition stands. They'd like to put them in boxes. ...Even something like Greenpeace who do take direct action...I think...the local authorities know how far Greenpeace will

> go.... When somebody like Plane Stupid comes along, they are an unknown entity. It keeps the authorities on their toes (Interview with John Stewart, 9 December 2011).

The uncertainty felt by government bodies increased the interest of the media. However, both Plane Stupid and Climate Camp had planned direct action tactics as a general-organizational domain strategy, not one specifically geared toward the campaign. For Climate Camp, it was 'one of the core principles of the camp from the very, very beginning' (Interview with Hannah Garcia, 16 July 2012). Their novelty was not by strategic choice.

One strategic tactical decision that was made during the campaign and that did influence outcomes was the decision by HACAN and NoTRAG to commission a report on the economic necessity of the third runway. This report was later used by the Conservative Party to defend its position to oppose the runway, an important factor in the policy outcome. Such a strategic decision was made after gauging the available opportunities and understanding the needs of various actors that the campaign attempted to influence (Interview with John Stewart, 9 December 2011).

The campaign's overall approach regarding tactics was one developed strategically alongside intra-campaign relations. The loose coalition that Stewart worked to develop as part of his strategic leadership was centered on the philosophy 'unity of purpose, diversity of tactics' rather than on a particular tactic or political perspective (Interview with John Stewart, 9 December 2011). This philosophy allowed each group to engage in tactics in which it had competence (Interview with Hannah Garcia, 16 July 2012; also see Ganz, 2004). In addition, this allowed the campaign to pursue both inside and outside tracks simultaneously (Interview with John Stewart, 9 December 2011).

Green Investment Bank

In the case of the GIB, campaigners felt that 'what's most likely to persuade political parties' was lobbying by campaigners, rather than the public (Interview with David Powell, 18 October 2012). The campaign was 'focused on mobilizing MPs and some key industry stakeholders... who've signed on to some of the advocacy materials.... Basically, the key big investors, their voice would be listened to in this' (Interview with David Holyoake, 25 October 2012). Despite an emphasis on elite-level lobbying and investment-sector support, there was some small level of public lobbying in a few stages of the campaign (Interview with David Powell, 18 October 2012). When the bill was in the Lords, mobilization ceased entirely. This was strategically assessed. It was determined that public lobbying of the House of Lords was an ineffective strategy due to the fact that peers lack a geographic constituency, have the role of policy scrutinizers, and often do not have publicly available contact information (Interview with David Powell, 18 October 2012).

Despite this assessment, research utilizing interviews with peers found that '[m]ost peers in the absence of their own researchers, policy officers and administrative support, relied on [NGO lobbying efforts] as a key source of information on how policies and legislation will affect the general public or stakeholders. Such interaction acts as a "nudge", encouraging them to pay attention to various and sometimes specialist areas of legislation' (Foreman, 2012, 27). Peers also have a sense of public accountability 'to listen to what people say, and to use stakeholders' (Baroness Thornton, in Foreman, 2012, 27). Such a view among peers has in part been affected by previous public mobilization campaigns (Foreman, 2012, 31). A tactic of public lobbying was, therefore, a possible strategy with some possibility of succeeding. However, campaigners determined the nature of the campaign limited public enthusiasm and ruled it out. The inclusion of the tactic could have made some difference, but the strategists were probably right in their assessment of the public's unwillingness to engage with the campaign. It would have taken an additional campaign to convince the public of the virtues of the GIB, something deemed unnecessary as policymakers had already approved the policy.

Conclusion

Despite Gamson's early contribution, strategy is still under-researched in the outcomes literature (Ganz, 2004; Jasper, 1997). The findings presented here can help develop understanding of the role of strategy in social movement outcomes. Three key findings can be pointed out regarding the influence of strategic decision-making and agency in the campaigns.

First, *tactics were largely based on previous strategic choices* made at the general-organizational domain or in the campaign domain. For example, while the principle 'unity of purpose, diversity of tactics' was important for the campaign against the third runway, it was largely determined within the extension decision that predated more specific tactical choices. Nevertheless, some tactical decisions did seem to play some role in outcomes. FoE's choice of using celebrities appeared to generate additional enthusiasm as well as policymaker support. In addition,

HACAN and NoTRAG's strategic decision to commission a financial report on the third runway eventually helped the Conservative Party oppose the third runway, or at least reduce the costs of doing so. While these can be perceived as relatively small decisions, they had disproportionate effects on outcomes. The tactics that proved useful were the result of decisions made by leaders of the campaigns. This brings us to the second important finding.

Campaign leaders were crucial in making important strategic decisions, supporting Ganz's argument that strategic leadership plays an important role in outcomes. 'Strategic thinking is...based on ways leaders learn to reflect on the past, attend to the present, and anticipate the future' (Ganz, 2004, 180). John Stewart exemplified this type of leader. Stewart learned lessons from his past experience as an anti-roads activist, including the avoidance of the consultation process in favor of a more outsider-track approach (Interview with John Stewart, 9 December 2011; also see Saunders, 2004, 232). He was able to gauge important opportunities and obstacles during the campaign, realizing that the Conservative Party's rebranding attempt gave the campaign leverage but that the economic argument was still one that needed to be tackled in order to overcome resistance inside the party. He was also able to foresee future problems and thus avoided the NIMBY label from the outset, establishing a national network of groups opposed to airport expansion and helping to introduce the climate change frame through his extension strategy. While it was not discussed here, the Big Ask campaign was initiated by a strategic decision. Tony Juniper and other strategic leaders of FoE noticed a 'dramatic repositioning of the climate change issue coming from several directions at once', read it as an opportunity, and 'decided to change strategy,... dropped some of the old campaigns we were running...[and] rolled a lot of those resources into a new climate campaign' (Interview with Tony Juniper, 18 September 2014).

Third, *campaign leaders were responsible for developing an overarching, holistic strategy that answered each of the strategic questions*. For the Big Ask campaign, this strategy was to form a campaign run by Friends of the Earth that would apply pressure on policymakers through public lobbying. This strategy answered extension and relational questions as a package. For the campaign against the third runway, the decision to extend the campaign widely, and, therefore, loosely, answered campaign domain questions regarding tactics ('unity of purpose, diversity of tactics'), leaving individual groups to answer strategic questions for themselves but hoping to generate enough diversity and unity to pressure the government on a wide variety of fronts. The Green Investment

Bank campaign's strategy was rooted in providing the government with a persuasive argument during a time a policy window was open. This strategic decision impacted the answers to most other strategic questions. Smaller strategic choices played some role in outcomes, but these overarching choices made the most impact in the campaigns.

How did these strategic choices influence policy outcomes given that political opportunities also play a role? For the Big Ask campaign, political opportunities certainly got the ball rolling, but campaigning was important for generating large levels of cross-party support. The lobbying efforts were very much a product of campaigners' strategies to mobilize the public to both persuade and pressure policymakers. Other strategic choices may have resulted in the government failing to adopt the Climate Change Act, although this may not have been the case given the large levels of initial support for the policy by policymakers. It seems more certain that important amendments that strengthened the policy may not have been achieved had strategic decisions been different.

In the campaign against the third runway, strategy was crucial in exploiting political opportunities which may not have manifested into policy change without strategic choices. Developing a broad and loose coalition allowed for new frames to be attached to the campaign (especially climate change) and new approaches to be taken (especially direct action campaigning, which was able to make headlines). The greening of the Conservative Party was also an opportunity that was exploited by campaigners who reduced the costs for the Conservative Party leadership to oppose the runway.

The Green Investment Bank is an interesting case because its strategy was tailored to an open policy window, which subsequently closed during the campaign. Strategy was less important in this case because few promising options were provided to campaigners. Persuasion was an appropriate means of engaging policymakers during an open policy window, and pressure would have been difficult to develop during a banking crisis. Therefore, the Green Investment Bank is an example of the limits of strategy when opportunities are not ripe. The extent to which the GIB was a positive outcome was largely due to political opportunities.

Political opportunities set the territory for strategic decision making. Campaigns cannot win even with a perfect strategy if the time for change is not right. However, political opportunities and other structural variables do not explain outcomes either. Strategic choices play a noticeable role in policy outcomes. This means that those strategic decisions, and the leaders that make them, are important to investigate if we want to gain deeper insight into social movement outcomes.

7
Mechanisms for Policy Change

Mechanisms represent causal processes that produce an effect (Hedström and Ylikoski, 2010). By exploring mechanisms of movement outcomes, we can develop an understanding of *how* social movements influence policy. While the use of particular mechanisms can be the result of strategic choices by campaigners (see Chapter 6), we are not interested in the influence a choice had on the campaign but which mechanisms the campaign used and how effective those mechanisms were in obtaining a policy outcome. This can help us understand the paths to social movements' outcomes and the processes that facilitate policy change.

It is not uncommon for movements and campaigns to apply a variety of mechanisms at the same time or during the course of campaigning. Therefore, testing the ability of multiple mechanisms to produce outcomes within a single campaign is possible and desirable. In order to do so, we need to examine all the relevant mechanisms so as not to miss a crucial piece of the puzzle.

Kolb (2007), borrowing from the previous literature (for example, Andrews, 2001; Knopf, 1998; McAdam and Su, 2002), explicitly describes five mechanisms pertaining to policy change. These mechanisms (disruption, public preference, political access, the judicial mechanism, and international politics) concern the channels of influence on the political system and the ways in which outside forces can use these channels to affect policies. The following is a description of each mechanism followed by a brief analysis of the role that mechanism played in each case. Note that not all mechanisms were used in the cases.

Disruption mechanism

The *disruptive mechanism* argues that direct action, protests, or riots that interfere with important functions of society, or with the state directly, will result in positive outcomes because policymakers and the state desire a return to normalcy. The state, attempting to restore order, could either accede to demands or launch a repressive counterattack that the movement would then have to overcome. It is argued that if disruption can continue for long enough and outlast repression by the state, the movement is likely to win policy change. Piven and Cloward (1977) have argued that this mechanism is superior for producing outcomes when compared with mechanisms that relied on organizations and organization building in cases where communities have little power or are lower class. Busby's (2010) discussion of the role of physical force on influencing policy can also fall under this mechanism. This coercive force becomes a form of disruption and disquiet. Militant organizations and 'urban guerrillas' have been known to utilize such a mechanism in order to obtain specific policy demands.

Looking at our three cases, the disruption mechanism was only attempted in the campaign against the third runway. While many 'direct action' events, including banner drops, barricades, and flash mobs, occurred in this and other campaigns, these actions were intended mostly to attract media and public attention, more closely resembling tactics oriented toward the public preference mechanism (see below). While Climate Camp could have attempted to disrupt air traffic at Heathrow, a potentiality that worried the government, they agreed not to and no activists worked independently to do so. HACAN and NoTRAG's flash mobs explicitly avoided disrupting the airport but tried to generate media coverage with the threat of disruption (Stewart, 2010). This was similar to instances of protest in the other cases.

The tactic in which disruption did play a role was the Airplot where Greenpeace attempted to use 'legal **monkeywrenching**' to delay construction and increase the costs of the third runway. It would force the government to issue compulsory purchase orders for thousands of people. While Transport Secretary Geoff Hoon indicated that a delay and costs would be minimal (Sawer, 2009), Greenpeace had additional plans for an 'impenetrable fortress' to be developed on the land and be used by protesters to prevent construction by resisting being forcibly removed from the site. However, while it was probable that this would have resulted in increased costs and some delays, neither the compulsory purchase orders nor the fortress were tested. Therefore, it is difficult to test the actual results of disruption. While the policy on the third runway changed, disruption had not played a role in this change. It did not appear that the threat of disruption created a change either. This does not mean that disruption is not a useful mechanism, or that a similar result would not have occurred if disruption were tested, but that disruption was not a necessary condition for the campaign to achieve its outcomes.

Public preference mechanism

The *public preference mechanism* assumes that politicians desire public approval and are, therefore, interested in the public's positions on policies. If a movement can influence public opinion or issue salience, it can indirectly influence policymakers, employing the public preference mechanism. In democracies, this mechanism suggests that politicians, as elected officials, are accountable to their constituency to a significant degree. In order to get re-elected, politicians must consider the public's preferences in their own policy decisions. If movements can raise the salience of an issue and bring public opinion onto their side, then policymakers may support that issue in order to gain favor with their constituency. This mechanism views that a movement's ability to influence policy occurs indirectly via the public. A movement's ability to influence policy using this mechanism can depend on government openness and instruments of direct democracy because these affect the flow of information and power from the public to policymakers.

Often in the outcomes literature 'the public' is perceived in broad terms and is measured using data that samples the whole population within a policy scale (Burstein, 1991; Burstein and Linton, 2002; Giugni, 1999). That is, when looking at national policy, research often investigates a campaign's ability to affect national public opinion by looking at national opinion polls. In the present cases, opinion polls concluded that, while public opinion was high regarding government action on climate change, issue salience was low regarding how important government action on climate change was relative to other issues. When it came to how this affected voting, climate change did not seem to play a major role. Nevertheless, we cannot rule out the public preference mechanism entirely. Rather, we have to a) consider the mechanism within a context of structural and dynamic political opportunities and b) question the importance of polls in transmitting public opinion to policymakers.

Under the public preference mechanism, the electorate becomes important because it provides incentives and disincentives to policymakers for taking action on policy. Electoral calculations made by a policymaker, candidate, or political party, which states that the particular demographic is important for winning the constituency in either the short or long term, can adjust their perspective on any given policy. These calculations would be made based on the structural processes that make up the electoral system and dynamic processes that are specific to a particular election cycle. For example, the first-past-the-post system in each constituency means that only some constituencies will be battlegrounds. Which constituencies are battlegrounds will be context-specific. By looking at the battleground constituencies, political parties can maneuver in order to attract support in those constituencies without being interested in national opinion on a given issue as long as any maneuvers do not reduce support in other constituencies. While polls can help policymakers determine public opinion even within a particular constituency or demographic, policymakers can use a wide variety of tools to gauge public opinion, including protest attendance, levels of lobbying by their constituents, and the media.

If we assume that only national opinions are important for national policymakers and that those opinions can only be gauged via national opinion polls, we may be missing crucial piece of the movement outcomes puzzle.

Climate Change Act

Prior to Cameron becoming leader of the Conservative Party, the environment was not a contested issue in the political arena (Carter, 2008). By the end of the Big Ask campaign, climate change was a political issue because it represented a turning point for the Conservatives, generating political competition around the issue (Carter and Jacobs, 2013). This was not the result of changes to large-scale public opinion but a re-evaluation of the perceived importance of opinions of particular demographics. The fact that young people both did not vote for the Conservative Party and had strong views about climate change became important. The Conservative Party sought out these voters through a rebranding campaign that also involved adopting environmental policies that the party had previously opposed.

Friends of the Earth, a trusted ENGO, proposed a national climate change policy. In doing so, they were able to attract the young people that the Conservative Party, under Cameron's leadership, sought to attract. This led the Conservatives to support the issue. While this was important, the attraction of young people to the issue of climate change was not something Friends of the Earth accomplished in this campaign. Another important part of the campaign was the signatories to the EDM, which helped put pressure on the government by showing it had large-scale cross-party support (Interview with Tony Juniper, 18 September 2014). While many of these signatories also did not need further persuading, others appeared to be persuaded by FoE's efforts. It is possible that rather than judging public opinion and issue salience based on polling data, policymakers themselves interpret lobbying efforts as reflecting the desires of their constituency. Campaigning efforts and the actions of nearly 200,000 individuals could have been interpreted by policymakers as a sign of strong public opinion and issue salience around the Climate Change Act. This could have been the perception not only of individual MPs that were lobbied but also of the prime minister. According to Tony Juniper (Interview, 18 September 2014), Tony Blair told him during the course of the campaign: "until you get me the public support, I can't do anything." This suggests that the mechanism did play a role in the Climate Change Act.

Heathrow Third Runway

Many of the actions of the campaign against the third runway were directed toward attracting public attention, first at the local level and then jumping scale to the national level. Media was the primary means to attract attention, increase the salience of the issue, and affect public opinion on the third runway. Attaching the climate change frame to Heathrow was important in this regard, particularly as it resonated with a particular demographic that was being courted by the Conservative Party, namely younger voters (and perhaps residents near to the proposed runway who were in marginal constituencies). As stated above, policymakers may be seeking the favor of a particular demographic rather than a whole population. In this case, interest in young voters was not tied solely to marginal constituencies but across all constituencies. Therefore, campaigns that appealed to younger voters nationally provided more incentives for the Conservative Party.

The campaign against Heathrow expansion was able to generate media attention and effectively tie together the issue of climate change and airport expansion. The Conservative Party was seeking to shift its position on the environment in order to attract younger voters and combat negative images of the party. The media attention that was generated by the campaign around the third runway made it a nationally salient environmental policy particularly due to the climate change frame that the movement attached to it. Without such a campaign, it was unlikely that the third runway would be perceived as a climate change issue or reach national significance. The Conservatives decided to make a policy U-turn on the issue and in doing so marked a crucial point in achieving the policy outcome. Therefore, the public preference mechanism must be credited for much of the influence campaigners had on the policy.

Green Investment Bank

Although climate change was never a highly salient issue, it dropped in public salience as economic issues and austerity became the central political concern. One 2011 poll that mentioned the Green Investment Bank found that 44 percent of respondents supported canceling funding going to the bank (only 33 percent disagreed; YouGov and The Taxpayers' Alliance, 2011). Polls also found that public perceptions of banks were generally negative (see Asthana, 2010). It could be argued that the low level of public preference may have contributed to the bank not being granted immediate borrowing powers or other powers. However, it did not seem that general public preference had an enormous role in the GIB's *success*. While parties did compete on green issues prior to the election, which resulted in the GIB being adopted by both the Labour and Conservative parties, and this can be tied to vying for public support around the issue of the environment, the GIB campaign had no role in influencing this party competition or attracting public interest to environmentalism. It must also be said that, because of the economic climate, campaigners did not seek to use public preference as the central mechanism of the campaign. Their strategic assessment was that the GIB would become legislated and would be strengthened based on the support from the financial community and the financial arguments the campaign would develop.

Political access mechanism

The *political access mechanism* states that social movements can influence policies when movement actors are granted access to the formal political arena by being elected to office or being appointed to a political post, or when their 'beneficiaries' (see Gamson, 1975) gain voting rights or other political powers. Once deeper into the political system, social movement actors or beneficiaries can influence the political agenda in favor of their movement's position. Therefore, movements can seek to run candidates for office or campaign for obtaining further political power for their beneficiaries in order to help them influence policy in the future. Another avenue for political access is *conversion*, or the ability to persuade policymakers to truly alter their political position and internalize the movement's policy aims (Busby, 2010). This would create a new de facto movement member already in a position of power that is able to directly influence policy.[31] This differs from influencing a policymaker to change policy for the purposes of political expedience which would fall into one of the other mechanisms.

Climate Change Act

Bryony Worthington was a climate change campaigner for FoE who was called on to help draft the government's climate change bill. According to campaigners, this draft was inadequate in its emissions reduction target, it lacked annual targets, and it did not include emissions from international aviation and shipping. Worthington argued that they did not have enough time to prepare the draft bill. However, the speed at which the draft bill needed to be completed saved it from Treasury scrutiny. Again, the political opportunities available to Worthington to make the bill stronger had closed, while the threat of a cabinet reshuffle may have been the reason that the bill quickly moved past the Treasury who may have otherwise curtailed it. This suggests that the level of access reached by the movement is important, alongside the general political climate and the specific political context at the time. However, it could have been the case that the government's draft bill would have been weaker without the presence of Worthington.

Heathrow Third Runway

In the campaign against the third runway, some movement members were also politicians. John McDonnell is a clear example. He attempted to lobby Labour ministers and was a loud voice on the issue. However, he recounts that this was not particularly useful, and while some Labour ministers later tried to push for concessions, the leadership would not concede the third runway until after the party's electoral defeat. Like McDonnell, Conservative MP Zac Goldsmith was also a campaigner who tried to shift his party's position on the issue but did not view his role as particularly influential (Personal Correspondence with Zac Goldsmith, 15 September 2014). However, it is suggested that MP Justine Greening, an active movement member who attended and participated in protests and interacted closely with HACAN (Personal Correspondence with Justine Greening, 17 September 2014), was able to help move the Conservative Party on the issue. The Conservative Party U-turn played a major role and, while other forces were also significant in the party's decision, according to lead campaigner John Stewart, Greening played 'an important role' (Interview, 7 December 2013). This suggests that to be effective, movement members need to reach a political position proportionate to the importance of the issue on the political agenda.

Green Investment Bank

The GIB campaign did not have many activists engaged in the political process, although Worthington did participate in trying to push through amendments in the House of Lords. The campaign quickly got the ear of several politicians who were happy to support the bank and take on the campaign-proposed amendments. This cross-party support was very important for the success of the bank but represented political allies rather than converts or movement members. Without the support of others, Worthington was very limited in her capacity to get amendments approved. This reiterates the point that the level of access attained with regards to policymaking powers is important in successfully employing the political access mechanism.

Judicial mechanism

The *judicial mechanism* argues that by bringing a case to court, a social movement could change policy. A movement can launch a court case and hope that the court rules in favor of the movement's position and that the ruling then influences policy. Some campaigners prefer this approach when public opinion for the movement's cause is low. However, Rosenberg (1993) argued that, at least in the United States, the mechanism was less effective than some social movement leaders assumed. He showed that public opinion played a strong role in the implementation of judicial decisions, arguing that the role of the courts in social change is diminished because the judicial branch will only hear arguments based on pre-existing rights; it lacks independence as Congress can overturn their decisions; justices are influenced by other branches of government; and in practice justices rarely make unpopular decisions. Nevertheless, it is important to test this mechanism in order to see how such court challenges can be used for policy change, particularly outside the US context.

The judicial mechanism was used only in the campaign against the third runway, perhaps largely because the other two campaigns attempted to pass new legislation rather than change existing policy (although see the 2015 court ruling on greenhouse gas emissions in the Netherlands). In 2004, organizations representing campaigns against expansion in Luton, Heathrow, and Stansted airports coalesced to mount a legal challenge in opposition to the Aviation White Paper, which had proposed widespread airport expansion. This marked the first judicial challenge to a government white paper in UK history. Other lawsuits were also filed. These suits resulted in a largely favorable verdict in February 2005, but the verdict mostly regarded other airports in southeast England and only delayed the expansion process (HACAN ClearSkies, 2005a). How important was this delay? By December, a pre-budget report placed Heathrow back at the center of expansion plans (HM Treasury, 2005), expecting the third runway to be complete by 2017.

Another legal challenge in 2010 was a major victory, but it did not stop expansion. The verdict required the government to rethink the proposal considering emissions targets, which may have at least required additional concessions. However, the importance of this judicial challenge was not tested because the coalition government rejected the third runway when they formed a government. The verdict itself did not influence the parties' position on Heathrow expansion.

International politics mechanism

The *international politics mechanism* refers to pressure that can be brought upon policymakers by forces outside of their country and outside of the control of their state. A social movement can appeal to various outside forces in order to implement the mechanism. Kolb identifies four variants (2007, 90–1):

1. Leveraging global markets (for example, international boycotts)
2. Citing international commitments or pressuring international organizations
3. Increasing international awareness and utilizing international normative discourses
4. Changing the national political context by transnational campaigning or foreign support or scrutiny

One example of this mechanism at work in international climate change policy is when the head of the World Wildlife Fund's delegation to the Kyoto Protocol negotiations, Jennifer Morgan, hammered home the message of 'national honor, prestige, and reputation' when speaking with the Japanese government about ratifying the Kyoto Protocol. The desire to align with the international community was later cited as a critical reason for Japan's policy change (Busby, 2010, 144; also see the case of Canada, Busby, 2010, 146–7; cf. Keck and Sikkink, 1999; Tsutsui and Shin, 2008).

The nature of climate change makes the international politics mechanism more likely to be important, particularly when discussing national climate change mitigation policy. If high-carbon-emitting countries do not agree, individually or collectively, to reduce emissions, no amount of effort by the UK will prevent the significant effects of climate change from occurring. Therefore, it stands to reason that campaigners would employ an international politics mechanism. However, this was only true of the Climate Change Act. In the case of the third runway, international politics, specifically EU pollution regulations, were a delaying force that prohibited some of the initial plans for expansion. While campaigners reported on these delays and regulations, they did not pursue a strategy that involved international politics, perhaps realizing that the

EU regulations would come into play with or without their campaigning efforts. Regarding the Green Investment Bank, climate change was not a major frame in the campaign due to strategic choices that assessed the political context at the time. Therefore, there was less necessity for international politics to become a relevant factor.

Climate Change Act

As discussed in Chapter 7, the use of the international leadership frame regarding the Climate Change Act was far reaching, cross party, and central to justifying the legislation. The frame allowed the costs of implementing the Climate Change Act to be justified because of the precedent it would set (thereby reducing risk for other countries to participate in emissions reduction and providing a framework by which to legislate reduction targets) as well as increase international recognition. The possibility of other countries enacting their own climate change legislation could be used to combat arguments by those who were skeptical of unilateral action. The sheer number of occurrences of the concept of international leadership within the government's draft bill shows how important it felt the international political mechanism was for the act. However, there is a question of the extent to which campaigners had successfully utilized this frame rather than the frame being a defensive position by a government ready to adopt the act for other reasons. No 'smoking gun' evidence was available, but as both campaigners and the government used various other arguments for the Climate Change Act, it would be safe to assess the international politics mechanism as only moderately effective.

Conclusion

The results (see Table 7.1) show that the public preference mechanism played a key role in two of the three campaigns. In both the Big Ask campaign and the campaign against the third runway, it represented the mechanism that had the most impact. This directly ties into the argument that a policy window opened up around climate change and that campaigners needed to exploit the opportunity in order to achieve outcomes. In the case of the Climate Change Act, the campaign had successfully attracted a demographic around which party competition formed following a strategic decision by the Cameron leadership. The Big Ask campaign's strategy of working with celebrities helped to attract young people to the campaign and generate publicity. In the case of the third runway, campaigners also generated publicity alongside a new

Table 7.1 Effectiveness of mechanisms across cases

	Disruption Mechanism	**Public Preference Mechanism**	**Political Access Mechanism**	**Judicial Mechanism**	**International Political Mechanism**
Climate Change Act	–	Effective	Moderately effective	–	Moderately effective
Heathrow Third Runway	*Attempted but not tested*	Very effective	Effective	(1) Ineffective; (2) *Attempted but not tested*	–
Green Investment Bank	–	Ineffective	Ineffective	–	–

frame on climate change that turned a local environmental and community issue into a national issue. In doing so, both campaigns were able to obtain the support of the Conservative Party, which resulted in party competition that led to policy change.

In the case of the Green Investment Bank, policymakers adopted the idea of the GIB but failed to adequately equip it in legislation because campaigners had lost leverage due to the closure of the political opportunity. The mechanism that explains the outcomes of the GIB predated the GIB campaign. Where GIB campaigners were helpful was in developing a policy idea that was quickly adopted by political parties. While the public preference mechanism explains the GIB case, it cannot be attributed to the campaign itself but only to the broader climate change movement that worked to increase salience and public opinion around the issue, particularly of younger people.

8
Conclusion

This book set out to better understand the role of social movements and civil society in policy change, looking in particular at a social, political, and environmental problem whose full impact we cannot predict: climate change. As social movements often seek to influence policy, it is important to investigate *what* specific elements of the policy process movements influence in order to develop an understanding as to where strengths and opportunities lie. Additional opportunities can be present in dynamic political processes that open and close policy windows. By exploring these political opportunities, I also sought to better understand *when* social movements were better able to influence policy. Policy change, even change within specific components of the policy process, can be approached in a variety of ways. Social movements can use a variety of mechanisms to attempt policy change, so it was important to understand which mechanisms were able to influence policy or, to put it simply, to investigate *how* social movements can change policy. Part of the answer to the *how* question requires an investigation into the powers of agency and the role of strategic decisions that movement leaders make. This, too, was investigated.

Each of these questions can be asked of social movements individually or collectively. I undertook a study of three cases within the climate change movement in order to better answer each of the above questions. The cases had important similarities and differences that deepened our understanding of social movement outcomes. First, the three campaigns addressed diverse issues within a broader climate change frame, allowing us to see a range of impacts across the policy process. Second, the cases spanned the period prior to and just following a policy window. At the same time, none of the three cases perfectly overlapped, so possible changes in the political or social context could have influenced the campaigns and their outcomes, and potential cumulative outcomes could be assessed. Third, the cases were contained within the same movement, but each case revolved around a different movement organization, allowing us to gauge the impact of strategic leadership as well as a more nuanced understanding of outcome mechanisms.

Why examine the climate change movement? The climate change movement is young and made up of a diverse range of actors whose climate concerns are often linked with other problems. Climate change activism first arose during international negotiations for a climate change treaty, with activists forming various groups that mostly focused on lobbying efforts and were institutionally tied to Climate Action Network. As discussed in Chapter 2, these campaigning efforts largely failed to advance a framework for greenhouse gas reductions as they were unable to adjust political calculations of key nation states in the negotiations process. Organizations concerned with the consequences of climate change began rethinking their approach, and national groups started to focus on influencing national policy as well as pressuring their governments to do more internationally. The UK became a prime example as organizations sensed a political opportunity following strong statements on climate change by Prime Minister Tony Blair. These organizations had loosely networked into a movement and launched a variety of campaigns. These included the Big Ask campaign that called on the government to legislate an annual 3 percent cut of greenhouse gas emissions that would result in a reduction of 80 percent by 2050; the campaign against a third runway at London's Heathrow Airport that attempted to curb the expansion of carbon-intensive infrastructure; and the Green Investment Bank campaign that advocated for the establishment of a government-funded bank that would invest in renewable energy, energy efficiency measures, and low-carbon technologies. These three cases were examined in detail to better understand social movement outcomes.

Findings

Political opportunities

The analysis presented in the book, particularly in Chapter 5, shows that a policy window was crucial for policy change in the three campaigns. Party competition around the issue of climate change, following the election of David Cameron as the new Conservative Party leader and his subsequent assessment that going green would boost the party's image, led to the opening of a climate change policy window. This window provided campaigners with a chance to push forward their favored policies or reframe concerns in climate change terms. The Big Ask campaign had already ignited prior to this party competition as Tony Juniper and other strategic leaders within Friends of the Earth had seen an opening in Prime Minister Blair's strong remarks about climate change (Interview with Tony Juniper, 18 September 2014). Nevertheless, the campaign significantly benefited from the greening of the Conservatives, who increased party competition, eventually leading the Labour government to pass the Climate Change Act 2008. Likewise, the campaign against a proposed third runway at Heathrow had experienced a similar advantage, although the Labour Party failed to out-green the Conservatives on the issue and did not oppose the third runway until after the 2010 election, which resulted in a Conservative-Liberal Democrat coalition government. The end of the election cycle, along with increased concerns over the economy, led to the closure of the climate change policy window, at which point campaigners were unable to make progress regarding the Green Investment Bank that was promised by the government. While a Green Investment Bank was established as an institution, it was not given borrowing powers that would have allowed it to leverage funds, effectively making it a green fund rather than a bank.

While it was clear that dynamic political opportunities were crucial, these movement outcomes were not inevitable. The Climate Change Act required public support before Tony Blair could support it (Interview with Tony Juniper, 18 September 2014). Even when it was accepted by the Blair government, it lacked some crucial features that required additional campaigning to resolve. The campaign against Heathrow expansion needed to raise the political stakes for policymakers to take strong positions on the matter. The issue was originally confined to local concerns that could be labeled NIMBY. Climate change activists were crucial in bringing the issue of airport expansion to the national and international arena, creating a wider environmental argument that policymakers could compete over. The idea for a Green Investment Bank had to be created and had to have support of businesses before it could be adopted by the government. Campaigners were instrumental in achieving both of those aims. Their absence may have meant little change to the funding and financing of climate change mitigation projects.

Strategy and leadership

Strategic decisions were found to be very important to the success of these campaigns (see Chapter 6). Again, Tony Juniper and other FoE staff had the foresight to see Tony Blair's statements about the climate change as an opportunity, leading FoE to divert resources and organizational focus on a new climate change policy. John Stewart of HACAN ClearSkies, the organization at the heart of the campaign against Heathrow expansion, was vital to bringing in greater participation to the campaign, providing policymakers with appropriate arguments and building a network of activists and organizations that would broaden frames used to oppose a third runway. Ed Matthew, head of Transform UK, played an important role in initiating a campaign for the Green Investment Bank and establishing a coalition that would provide the arguments for political parties to adopt the idea. He strategically assessed the need for all three political parties to get on board with the bank to ensure that it would be taken on by the parties that formed a government following the 2010 election.

These examples demonstrate the important role that leadership plays in assessing the information they received and integrating it into their decision-making processes along with the lessons learned from previous experiences. These leaders exploited political opportunities and created strategies that answered bundles of strategic questions, providing an overarching trajectory for the campaigns.

Policy outcomes

In order to assess outcomes, the Policy Outcomes Model was created. The model breaks down into distinct components the possible areas of influence a movement can have on a policy. The campaigns were able to make significant progress in influencing policy, particularly at the policy consideration stage. This highlights how, once a policy window is opened, policymakers seek solutions to policy problems, and civil society and interest groups can interpose their ideas at this stage. However, each time policymakers acted on these considerations, they fell short of the standards set by the campaign. The government's draft of the Climate Change Bill did not include the campaign's recommended emissions reduction target and failed to include emissions from international aviation and shipping. When pressured by a Conservative Party U-turn on the issue of Heathrow expansion, the Labour Party had an internal dispute which resulted in concessions being made in a direction favorable to campaigners, but Labour still supported a third runway. While the coalition government agreed to create the Green Investment Bank, their version was very weak on all counts: it was unsure how green it would be, it lacked sufficient levels of investment according to campaigners, and it was not technically a bank because it could not borrow. In the more successful cases, we can see that having impact on the political action component of the model helped to strengthen the policies, but it was easier to have such an impact when political opportunities were open. In the case of the Green Investment Bank, the campaign lost contact with the

Treasury when the policy window closed, making it nearly impossible to have the level of influence it had previously.

Mechanisms

The use of mechanisms is based on the nature of the campaign as well as the strategic choices of campaign leaders. In investigating the effectiveness of various mechanisms, the role of political opportunities was strongly supported. Mechanisms were most effective when they corresponded with available political opportunities. In the case of the GIB campaign, political opportunities seemed to determine its results. No mechanism was shown to be effective within the campaign itself, although the opening of the policy window and its duration were likely a result of more long-term campaigning by the climate change movement. Once the policy window was opened, movement organizations could propose a policy solution and have the ear of policymakers. However, the GIB, when legislated, was not a strong policy, as both opportunities and access for the campaign were shut. Where campaigners were more successful, the public preference mechanism proved especially effective. By attracting public attention, campaigners were able to leverage the perceived voting power of that public and influence policymakers' decisions. In the Big Ask campaign, large numbers of signatures on petitions and postcards sent to MPs created a perception of solid voter support for action. In addition, the campaign attracted a lot of young supporters who were also a desired voting bloc for the Conservative Party. In the campaign against the third runway, public preference was molded by extending the frame of the campaign from local environmental and community issues to the issue of climate change, coupled with the threat of disruption and direct action tactics, which facilitated international news coverage.

Implications

The research and findings of this book have several implications for understanding social movement outcomes regarding the a) defense of certain theories, b) novel findings, and c) overall theoretical contributions.

Supporting theoretical arguments

A consistent finding in the analysis presented here supports the political process model of social movement outcomes. While political opportunities and other structural variables do not explain *all* outcomes, it is certainly the case that a policy window opened when parties began to compete around the issue of climate change and that this window appeared to be a necessary but insufficient condition for movement outcomes. At the same time, leadership and strategic decision making within the context of an open policy window was also shown to be highly important. This is particularly noteworthy in the case of the third runway opposition, where resources were more limited. Following Ganz's argument, leadership proved a valuable resource in itself and was able to direct the campaign to achieve policy change through its strategy.

Finally, the effectiveness of the public preference mechanism also supports the argument, as defended by Burstein (1999; 2003), that public opinion and issue salience matter in policymaking and that movements are more effective when they can influence these factors, supporting a broader democratic theory of governance (Burstein, 1998), but one that highlights the opinions of key demographics and constituencies rather than the voting public writ large.

New findings

A majority of the findings presented here are new as little research on the outcomes of the climate change movement has been conducted. However, I focus here on novel theoretical contributions that can be applied in further social movement outcomes research. While most research focuses on larger-scale political opportunities, often tied to structural variables, it is important to note that micro-political opportunities have contributed to movement outcomes. These micro-political opportunities represent small openings that can, if acted upon, be used to influence campaign outcomes.

Cumulative outcomes can also act as a useful tool in both understanding movement outcomes over a range of campaigns and in prolonging the duration of a policy window. Cumulative outcomes represent the use of previous movement impacts to gain leverage in subsequent campaigns. The Climate Change Act, a movement outcome, was used in both the campaign to oppose the third runway and the GIB campaign because it represented a large-scale, long-term policy that set out to reduce greenhouse gases. Such a policy enables other campaigners seeking to reduce emissions to use the act as a way of justifying their demands or shaming government into action. This suggests that once opened, policy windows can be prolonged if policies are passed that provide additional opportunities for campaigning and have a long-term or broad impact on policymaking.

Theoretical developments

Some of the findings presented in this book lead us to consider wider theoretical developments. While previous studies have often examined public opinion and issue salience levels of the whole of the population, these factors did not correspond with policy change in the cases presented here. Climate change was consistently seen by the public as a problem, but it had a low level of salience. This did not seem to influence the policy window or the passage of specific climate policies. Nevertheless, the public preference mechanism did appear to play a significant role in two of the three campaigns. How can we explain this? How these concepts are operationalized and measured appears to be important. The theoretical underpinning of the public preference mechanism suggests that policymakers are concerned about electoral victories and will structure their policy agendas accordingly. If we consider the electoral institutions in countries such as the United States and the United Kingdom, we will find that elections for control over the executive or legislative branches of government are not won by national majorities but by pluralities in each constituency. In any given election, only certain constituencies can be considered electoral battlegrounds (for example, marginal constituencies or swing states). If policymakers make calculations along these lines, it makes little sense for researchers to look exclusively at nationwide opinion polls.

Dynamic political opportunities and micro-political opportunities require us to focus on the details of cases and prefer in-depth over broad analysis when investigating the role of opportunities because they are too fine to see when examining a wide array of cases or a long time span. Similar to a more nuanced approach to examining the role of public opinion and the public preference mechanism, a nuanced understanding of opportunities should be explored when examining particular cases, as these opportunities can cause a ripple effect of change. Some analyses that preferred broad examinations of movement outcomes and ignored such dynamic and micro-political opportunities while also reporting a lack of impact by social movements (e.g., Giugni 2004) may have missed important moments in which campaigners were able to react to smaller changes in the political and social context of their campaigns and create policy change.

On strengthening subsequent research

While this book has examined a relatively recent movement in novel and detailed ways in order to shed light on the subject of movement outcomes, it certainly is not without its faults, and its findings can be strengthened by further research. Subsequent research may learn not only from the present findings but also from gaps in the research.

We know that broad survey samples did not find public opinion or issue salience to explain the greening of the Conservative Party. The party's rationale for such a move may not be tied to even detailed survey indicators but to other factors. Therefore, an investigation into the precise factors political parties use to make decisions on demographics to court would go a considerable way into improving understanding of how policy windows open, how party competition can occur, and (potentially) how social movements can influence these factors and play a role in opening policy windows.

While we do not know the precise reasons for the greening of the Conservative Party, it is clear that it wanted to appeal to youth and move away from the party's 'nasty' image. Therefore, it is likely that the issue of climate change was prominent enough in the mind of young voters in part because of the work movement organizations had been doing prior to the three campaigns (Interview with Tony Juniper, 18 September 2014). However, assessing this impact was outside the scope of this book, and it deserves sufficient space and analysis. In conducting such research, we can develop a more long-term picture of movement outcomes and help to identify additional pathways to policy change.

Why the Conservatives picked the environmental issue to rebrand their party is another question left unanswered in this book. Did the climate change movement play a role? Was the issue as compatible with conservative politics as Cameron had stated (Cameron, 2006)? Was it politically palatable because it was 'pain-free' (Toynbee, 2008) and did not rely on advancing strong policy at the expense of other Conservative Party values? This, too, requires further investigation as it may be of additional use in predicting or promoting the appearance of policy windows.

Finally, while this study attempted to single out dynamic political opportunities by studying several cases within a single country, it can also be useful to understanding the role of political opportunity structures. In order to do this, researchers can undertake similar studies of cases within different countries and compare and contrast the results. In this way, the literature can amass knowledge that is both wide and deep.

Notes

1 Other climate researchers disagreed with the assessment and called for policymakers to wait for additional research (Christianson, 1999, 199–201).

2 The conference's closing statement called for a comprehensive and holistic international solution that was to be the subject of the follow-up International Meeting of Legal and Policy Experts on the Protection of the Atmosphere, again hosted by the Canadian government. However, the meeting failed to result in any such solution, being opposed to by executive director of UNEP, Mostafa Tolba, fearing lack of support from the United States (Macdonald and Smith, 1999–2000, 110).

3 The issue included an article on climate change already warning that 'overreliance on fossil fuels could produce an increasingly uncertain and potentially bleak future' (Lemonick, 1989; Dessler and Parson, 2010, 23–4).

4 For example, the UK was able to block EC attempts to establish an international environmental organization backed by the International Court of Justice (Cass, 2006, 25).

5 WCC-2 followed from a 1979 meeting in Geneva hosted by a variety of international organizations (particularly WMO) and attended by scientists from over 50 countries which developed scientific working groups and established the World Climate Programme.

6 While the INC prepared the climate change framework, the UNFCCC represented only one part of the Earth Summit. Preparation for the summit occurred in Preparatory Committee (PrepCom) meetings, which were restrictive to NGOs. The initial wording for the PrepCom implied that NGOs would not be involved; NGOs that would be allowed to participate were initially circumscribed by a restrictive definition (Willetts, 1996, 71), and participation at meetings was curtailed to speaking 'with the consent' of others present. The wording explicitly stated that NGOs would not have 'any negotiating role' or the right to issue statements as official documents (Willetts, 1996, 75). Although eventually access became granted to a large number and variety of NGOs, their role was still relegated largely to the sidelines (Van Rooy, 1997, 100–1).

7 CAN Southeast Asia was established in January 1992; CAN Latin American was present since 1995 under a different name.

8 Cooperation provided an opportunity for Northern NGOs to express the points highlighted by their Southern counterparts (Rahman and Roncerel, 1994, 248–9). For more on North-South NGO relations in the build-up to the Earth Summit, see Khor, 2012.

9 This included flexibility mechanisms (Rahman and Roncerel, 1994, 266–9) which some, including CAN Europe, accepted alongside greater regulation of global emissions trading (Chasek *et al*, 1998, 12).

10 Another option for NGOs was to sit in on, but not participate in, the formal sessions comprised of speeches by delegates (Van Rooy, 1997, 100–1).

Formal and informal conversations between delegations and NGOs were found to play significant roles in influencing delegations' positions on small but important components of the conventions (Van Rooy, 1997, 99–101). Fourteen countries even had representatives from NGOs as members of their delegation (Van Rooy, 1997, 101). This level of access was difficult to attain, however, and with formal participation of NGOs during the summit curtailed, some NGOs focused efforts on publicizing the issues and educating the public in their own countries by distributing literature and holding demonstrations (see for example, Van Rooy, 1997, 102).

11 Additional targets were schedules to be negotiated at COP 3, but many questions still remained: How would emission reduction targets be distributed? Which greenhouse gases would be included? In preparation for COP 3, the Ad Hoc Group on the Berlin Mandate held sessions where these ideas were discussed. ENGOs played an important role in this process, with an average of 100 ENGO representatives attending each session (Corell and Betsill, 2001, 93). Once again, action among ENGOs was largely coordinated through CAN, which lobbied parties on a variety of issues (for example, Earth Negotiations Bulletin, no date b), disseminated information to negotiators, spoke with the media, and made at least one formal collective statement in each session (Corell and Betsill, 2001, 95, 98). However, ENGOs' access during much of the negotiations was symbolic. Time allocated to NGOs to speak during the session often occurred at the very beginning or end of meetings, and the allocation of only a single slot for all ENGOs meant that the statement would be of a general nature; in addition, NGOs were excluded from important informal discussions (Skjærseth and Skodvin, 169–70; Depledge, 2005, 221). Industry lobbyists were also present and sometimes outnumbered NGOs (Raustiala, 2001, 99). For more on the access and participation of NGOs to the climate change negotiations, see Depledge, 2005.

12 The environment commissioner at the time stated: 'We are fortunate to have a lot of activist NGOs to push nations along' (as quoted in Betsill, 2008, 54).

13 For more on the role of NGOs in this period, see Gulbrandsen and Andresen, 2004.

14 Like in Russia, environmentalists played an important role in Japan's ratification of the Kyoto Protocol. Japan ratified the protocol only after significant pressure from international NGOs who relied on framing the protocol as important for preserving Japan's reputation as a good international citizen (Busby, 2010).

15 National and international ENGOs also attempted to influence the negotiations by engaging in dialogue with policymakers from individual countries (see for example, Oberthür and Ott, 1999, 72). After Japan's offer to host COP 3 in Kyoto was approved, the Japanese delegation held meetings with international ENGOs to discuss their position (Oberthür and Ott, 1999, 54). In the US, ENGOs' attempts to lobby policymakers were overshadowed by the opposing efforts of the fossil fuel lobby. ENGOs organized public meetings in various cities to increase public support for the international negotiations and increase pressure on US policymakers, with similar events also held in Europe (Oberthür and Ott, 1999, 76). In Germany, the environmental movement was unable to substantially influence climate policy but had significant impact on public opinion (Skolnikoff, 1997) and the Federal Ministry of the

Environment, although they themselves lacked significant power to influence climate change policy (Sprinz and Weiß, 2001, 84). Within the European Union, environmentalists attempted to lobby for stronger commitments but were largely overshadowed by the lobbying strength of business groups who were more organized on the EU level and better funded (Sprinz and Weiß, 2001, 81). In India, the Centre for Science and Environment and the Tata Energy Research Institute helped the government to understand issues that arose during negotiations, but otherwise ENGOs in India are relatively weak (Sprinz and Weiß, 2001, 87–8). This showed that the strength of the ENGOs relative to business interests was important in determining the role national delegations played during the negotiations (see Sprinz and Weiß, 2001, 89).

16 Tickets sold out after only two minutes (Friends of the Earth, 2006a).
17 Prior to the launch of the Big Month, a trial run was attempted with MPs Kate Hoey and Neil Gerrard. Hoey stated that she would send a letter if more constituents asked her to but that she thought it would not be very effective; Gerrard was more supportive.
18 Its size was only surpassed in 2009 when demonstrators gathered during COP 15.
19 For London mayor Boris Johnson's retort, see Mulholland, 2008.
20 Not all Aldersgate Group members felt this way. A few were hesitant to support government investment in fear that it may 'disrupt a niche market that they occupy' (Interview with Peter Young, 8 November 2012).
21 This was particularly important as the campaigners' desired legislation would feature novel components, including having the GIB be a registered company under the Companies Act while having its own legislation and acting at arm's length from the government (Interview with David Holyoake, 25 October 2012).
22 Transform UK and Client Earth (2012), in the memorandum, had suggested that amendments to the bill should be addressed all together instead of separately in order to save time. This was rejected.
23 When discussing the cases presented here, we are looking at a period during which a policy window around climate change was open in the UK (Carter and Jacobs, 2013), meaning that the agenda was set for some policy options to be considered on the topic of climate change and the environment (see Chapter 5). It could be the case that climate change campaigners, or the environmental movement more broadly, may have had an influence on setting the agenda and opening the policy window, but that is beyond the scope of this book.
24 This represented 22 percent of Conservative MPs, 73 percent of Liberal Democrat MPs, and 30 percent of Labour MPs.
25 Because the days in between recesses were not divisible by seven, MPs that signed the EDM just prior to recesses were dropped to maintain consistency. Five MPs who signed the EDM during the time analyzed were, therefore, unaccounted for.
26 Here, marginal seats include the following labels under the Electoral Reform Society dataset (see Stoddard 2010): 'Marginal – a marginal seat held by a party other than Labour, including Con/LD seats, a few Con/Lab seats, SNP seats, LD seats, NI etc', 'Marginal2 – the real front line between Labour and

Conservative government', and a three-way marginal race between all three major parties.

27 It must be noted, however, that the issues with POS are not easily overcome by conceptual clarity, as some of the problems arise in the fine lines that distinguish one variable from being political or non-political and structural or non-structural.

28 Other micro-opportunities were present but had more influence on mobilization than on outcomes. For example, the Aviation White Paper lacked specifics on what would be expanded. According to campaigner John Stewart, the white paper produced 'one, rather perverse, advantage', 'that it envisaged so much expansion it provided a very clear focus for...groups to unite around' (2010b, 15; cf Rootes, 2013a, 102–3).

29 Transform UK had a total of ten press releases during the course of the GIB campaign. Two of these were excluded, one for being too brief and not including a position on the GIB and the other for not discussing the GIB, instead referring to a different Transform UK campaign. FoE web pages that were included in the content analysis were found using a search for 'green investment bank' on the FoE website and excluded financial reports, reports by other organizations, and broken links. In total, 73 FoE web pages were analyzed.

30 It must be mentioned that an important strategic decision was made regarding Climate Camp's decision to protest at Heathrow. 'Those of us who started Plane Stupid...argued strongly that [the next protest camp] should be at Heathrow, and as it turned out a consensus formed that it should be at Heathrow' (Interview with Joss Garman, 4 July 2012). The decision was not taken lightly, and a consensus did not form easily. Many campers were wary that their involvement would be perceived as targeting passengers and individual travelers, which Climate Camp wanted to avoid (Interview with Hannah Garcia, 16 July 2012). Another concern was that a camp at Heathrow would soften the power of subsequent camps: '[I]t's so iconic, it's just such a massive thing to try and fight off. What on earth would we do for an encore[?]' (Interview with Hannah Garcia, 16 July 2012). Eventually, the group agreed to hold the camp at Heathrow, but some felt upset that the location was pushed so strongly by Plane Stupid activists, who were seen as having a vested interest (Interview with Hannah Garcia, 16 July 2012).

31 One analysis found that conversion played a role in influencing a key policymaker to support the goals of the Jubilee 2000 movement, a religiously framed advocacy campaign attempting to cancel Third World debt (Busby, 2010).

Glossary

Campaign: A campaign can be defined as a series of events, actions, and efforts geared toward achieving a particular outcome within the remit of a social movement. While a social movement must be comprised of various networked actors, a campaign can be undertaken by a single organization or group that is otherwise networked as part of the broader movement.

Campaigners: Campaigners are the individuals, both those working independently or as part of organizations or groups, that engage in a campaign. See *Campaign.*

Civil disobedience: Civil disobedience consists of individual or collective action in the form of intentional law breaking, often but not always in a non-violent manner. Henry David Thoreau's essay 'Resistance to Civil Government' was later renamed 'Civil Disobedience', helping to spread the use of the term. Later figures such as Mohandas Gandhi and Martin Luther King, Jr. were proponents of civil disobedience, respectively using principled and tactical non-violent methods. Also see *Direct action.*

Cumulative outcomes: Cumulative outcomes are movement victories achieved, in part, using the gains of previous campaigning victories. This can include framing previous victories or exploiting new opportunities that previous outcomes were able to create. Also see *Outcomes.*

Direct action: Direct action has a range of meanings among social movement actors (Interview with Hannah Garcia, 16 July 2012), making the concept difficult to pin down. These meanings usually fall into two broad understandings of the term. 1) Direct action can be defined as a social movement strategy in which individuals, organizations, processes, and systems that are perceived to cause problems are directly confronted in order to right perceived wrongs in a way that prefigures an alternative society. This version of direct action is uninterested in lobbying policymakers but is focused on directly addressing movement or campaign concerns. 2) Another definition of direct action is action that includes confrontation or disruption for the means of achieving a political or social objective. These actions can be violent or non-violent and often include intentional law breaking. As campaigners usually used direct action to mean the latter definition, this is how the book utilizes the term. Also see *Civil disobedience.*

Dynamic political opportunities: Dynamic political opportunities refer to the possibilities and constraints that instable and variable features of a political context provide to social movements' efforts to mobilize and achieve outcomes. This includes the opening and closing of policy windows, the relative strength of a political party, the party competition around a policy area, the locations of key constituencies within a given electoral cycle, the appearance of an important minor political party, the occurrence of a meaningful political event, the relationship between the government or key policymakers and relevant interest groups, the political ideology of the government or key policymakers relative to their own political party, and others. Also see *Political opportunity structures*.

Flash mob: Flash mobs are gatherings of people organized in a decentralized manner who create a scene 'in the dual sense of a drama disrupting the normal flow of activities and a stage' (Gore, 2010). The first flash mob was reported to have occurred in 2003 but has roots in actions taken by Youth International Party members of the late 1960s. The appearance of flash mobs has grown with the widespread use of digital technologies and social media that allow for rapid and wide-reaching communication that facilitates decentralized planning.

Framing: A process of engaging with the construction of meaning. In this case, it regards the use of meaning creation by social movement actors or campaigners in an attempt to achieve outcomes or mobilization. Framing is a concept derived largely from the work of Erving Goffman. David A. Snow and Robert D. Benford took this work and developed it within the area of social movement research. They specified three core framing tasks: 'diagnostic framing' (framing the problem and its causes), 'prognostic framing' (framing the proposed solutions), and 'motivational framing' (framing the rationale for participation in collective action) (Benford and Snow, 2000). Frames are often viewed as part of the agency exhibited by social movement actors.

Inside track: The inside track is the approach of social movements to work within established means of lobbying policymakers and engaging in consultation and other institutional practices that allow communication between policymakers and stakeholders. This may be compared with an outside approach that uses more unconventional means of attempting to achieve political or social change. Also see *Outside track*.

Micro-political opportunity: Micro-political opportunities are small, dynamic changes within a context of a social movement campaign that can provide openings to social movements' mobilizing efforts and outcomes. Whereas dynamic political opportunities concern broader variable changes within the political context of a campaign, micro-political opportunities regard shorter-term changes that often occur at departmental or local constituency level. See *Dynamic political opportunities*.

Monkeywrenching: This refers to actions and tactics that involve sabotage or property damage, partially as a means of direct action (definition 1) or to increase costs of taking actions opposed by social movement actors. This term stemmed from the novel *The Monkey Wrench Gang* by Edward Abbey (based on the real group the Eco-Raiders) which depicted environmentalists sabotaging machinery in order to preserve wilderness. This novel inspired the group Earth First! and, therefore, monkeywrenching is often associated with parts of the environmental movement. See *Tactic, Direct action*.

NIMBY: Not in My Backyard, or NIMBY, is a term used derogatorily against local campaigners, usually campaigners opposing development. It is meant to associate that campaign with selfishness, aligning the campaign with disapproval of development only in one's own local area. NIMBY 'is not an analytical characterisation but a political label stuck by the impatient politician or policymaker upon those whose resistance spoils the grand schemes of the planners and architects of policy' (Rootes, 2007b).

Outcomes: Social movement outcomes, or simply outcomes, are changes within the political, economic, or social sphere that are desired and achieved by social movements. They represent the success of conscious attempts of bringing about change, which can be contrasted with unintended outcomes that also occur. For more, see Gamson, 1975; Guigni, 1998; Kolb, 2007. Also see *Unintended outcomes*.

Outside track: The outside track is the approach of social movements to work outside of established means. This includes protests and civil disobedience, among other more unconventional forms of political participation. This can be compared with an inside track approach, which favors official channels of lobbying. Also see *Inside track*.

Political opportunity structures: POS refers to the possibilities and constraints that stable features of a political system provide to social movements' mobilizing efforts and outcomes. This includes the strength of the executive branch of government, the type of electoral system, the availability of citizen-initiated referenda, the length of the electoral cycle, independence of the judiciary, and others. POS can be distinguished from dynamic political opportunities, which are more variable. For more, see Koopmans, 1999; Rootes, 1999. Also see *Dynamic political opportunities*.

Protest camp: A protest camp is the tactic of occupying outdoor sites for the purpose of claimsmaking. In the UK, contemporary social movement activists draw from examples of peace camps in the 1980s, such as Greenham Common and Faslane (Badcock and Johnston, 2009; Doherty, 2000). Eco-protest camps in the 1990s soon followed (Doherty, 2000). Camp for Climate Action represents a social movement group that utilizes the protest camp tactic. See *Tactics*.

Social movement: Diani, comparing definitions of social movements across various schools of thought, arrived at a consensual definition of social movements: 'a network of informal interactions between a plurality of individuals, groups and/or organizations, engaged in a political or cultural conflict, on the basis of a shared collective identity' (Diani, 1992, 13). Rootes adapted this definition, bringing in Eyerman and Jamison's view that collective identity could be as broad as 'knowledge interests'. He defines a movement as 'a loose, noninstitutionalised network of informal interactions that may include, as well as individuals and groups who have no organisational affiliation, organisations of varying degrees of formality, that are engaged in collective action motivated by shared identity or concern' (Rootes, 2007a, 610).

Social movement groups: Social movement groups can be understood as a body formed of informal sets of relations between individuals that share movement goals and are networked with other movement actors. These organizations are informal in that they lack a clear, rigid, and impersonal hierarchy, division of labor, or compartmentalization. Camp for Climate Action would be an example of a social movement group. See *Social movement organizations*.

Social movement organization: McCarthy and Zald (1977, 1218) define a social movement organization as a 'complex, or formal, organization which identifies its goals with the preferences of a social movement or a countermovement and attempts to implement those goals.' A formal organization is a body of 'coordinated and controlled activities that arise when work is embedded in complex networks of technical relations and boundary-spanning exchanges' (Meyer and Rowan, 1977, 340). They are composed of elements, such as departments and positions, which are rationalized and impersonal but 'are linked by explicit goals and policies that make up a rational theory of how, and to what end, activities are to be fitted together' (Meyer and Rowan, 1977, 342). Friends of the Earth would be an example of a social movement organization. See *Social movement groups*.

Strategy: Strategies represent 'the conceptual link we make between the targeting, timing, and tactics with which we mobilize and deploy resources and the outcomes we hope to achieve' (Ganz, 2004, 181). Strategies are developed in order to achieve a goal but are less specific than tactics and are composed of choices or decisions. Within campaigns, strategic questions can include issues of extension, relations, and tactics, as shown in Chapter 6. For more on social movement strategy, see Downey and Rohlinger, 2008; Jasper, 2004; Smithey, 2009. See *Tactics*.

Tactics: Tactics are specific actions or methods used in order to accomplish a task or achieve a goal. In social movements, new tactics are often created as a result of interactions between movement actors and opponents (Tarrow, 2011, 116). Tactics can be contrasted with strategies, which represent wider, less specific approaches to achieving goals. See *Strategy*.

Unintended outcomes: Unlike outcomes, which are conscious efforts by movements to influence change, unintended outcomes are changes resulting directly or indirectly from movement activity but that were not intended to occur. These can include biographical, cultural, and institutional changes, as well as political, economic, and social changes. For more, see Earl, 2006; Giugni, 1998, 2006; McAdam, 1989. See *Outcomes*.

References

Adams, W. M. (2001). *Green Development: Environment and Sustainability in the Third World*, London: Routledge.

AEF (2008). 'Aviation Emissions Cost Assessment 2008'. Available: <http://www.aef.org.uk/?p=290> Accessed 07 May 2012.

Agar, Jon (2011). 'Thatcher, Scientist', *Notes & Records: The Royal Society Journal of the History of Science*, doi:10.1098/rsnr.2010.0096.

Agarwal, Anil and Sunita Narain (2003 [1991]). *Global Warming in an Unequal World: A Case of Environmental Colonialism*, New Delhi: Centre for Science and Environment.

Agnone, Jon (2007). 'Amplifying Public Opinion: The Policy Impact of the U.S. Environmental Movement', *Social Forces*, 85(4): 1593–620.

Alley, Kelly D., Charles E. Faupel, and Conner Bailey (1995). 'The Historical Transformation of a Grassroots Environmental Group', *Human Organization*, 54(4): 410–16.

Almeida, Paul and Linda B. Stearns (1998) 'Political Opportunities and Local Grassroots Environmental Movements: The Case of Minamata', *Social Problems*, 45(1): 37–60.

Amenta, Edwin and Neal Caren (2006). 'The Legislative, Organizational, and Beneficiary Consequences of State-Oriented Challengers', in David A. Snow, Sarah A. Soule and Hanspeter Kriesi (eds.) *The Blackwell Companion to Social Movements*, Malden: Blackwell, pp. 461–88.

Andrews, Kenneth T. (1997). 'The Impacts of Social Movements on the Political Process: The Civil Rights Movements and Black Electoral Politics in Mississippi', *American Sociological Review*, 62(5): 800–19.

Andrews, Kenneth T. (2001). 'Social Movements and Policy Implementation: The Mississippi Civil Rights Movement and the War on Poverty, 1965 to 1971', *American Sociological Review*, 66: 71–95.

Ares, Elena (2008). *Climate Change Bill [HL]: Bill 97 of 2007–08, Research Paper 08/53*, House of Commons Library.

Asthana, Anushka (2010). 'New Poll Reveals Depth of Outrage at Bankers' Bonuses', *The Guardian*, 21 February [internet site]. Available: <http://www.theguardian.com/business/2010/feb/21/bank-bonuses-outrage-opinion-poll> Accessed 8 July 2014.

Badcock, Anna and Robert Johnston (2009). 'Placemaking Through Protest: An Archaeology of the Lees Cross and Endcliffe Protest Camp, Derbyshire, England', *Journal of the World Archaeological Congress*, 5(2): 306–22.

Barnett, Jon and W. Neil Adger (2003). 'Climate Dangers and Atoll Countries', *Climatic Change*, 61(3): 321–37.

Bartels, Larry M. (1996). 'Politicians and the Press. Who Leads, Who Follows?' *Annual Meeting of American Political Science Association*, San Francisco.

BBC (2010). 'Election 2010, National Results', [internet site]. Available: <http://news.bbc.co.uk/1/shared/election2010/results/default.stm> Accessed 8 July 2014.

Benford, Robert D. and David A. Snow (2000). 'Framing Processes and Social Movements: An Overview and Assessment', *Annual Review of Sociology*, 26: 611–39.

Betsill, Michele M. (2008). 'Environmental NGOs and the Kyoto Protocol Negotiations: 1995 to 1997', in Michele M. Betsill and Elisabeth Corell (eds.) *NGO Diplomacy: The Influence of Nongovernmental Organizations in International Environmental Negotiations*, Cambridge, Ma.: The MIT Press, pp. 43–66.

Blake, Joseph (2013). 'Back to the Heathrow Barricades as Government Gets Ready for an Airport U-turn', *The Guardian*, 16 December, [internet site]. Available: <www.theguardian.com/commentisfree/2013/dec/16/heathrow-airport-runways-u-turn> Accessed 8 July 2014.

Blundell, John (2008). *Margaret Thatcher: A Portrait of the Iron Lady*, New York: Algora.

Bodansky, Daniel (1994). 'Prologue to the Climate Change Convention', in Irving M. Mintzer and J. Amber Leonard (eds.) *Negotiating Climate Change: The Inside Story of the Rio Convention*, Cambridge: Cambridge University Press, pp. 45–74.

Bolin, Bert (2007). *A History of the Science and Politics of Climate Change: The Role of the Intergovernmental Panel on Climate Change*, Cambridge: Cambridge University Press.

Boon, Bart, Marc Davidson, Jasper Faber, Dagmar Nelissen, and Gerdien van de Vreede (2008). *The Economics of Heathrow Expansion: Final Report*, [internet site]. Available: <http://www.hacan.org.uk/resources/reports/4504.final.report.pdf> Accessed 5 July 2012.

Brown, Gordon (2007). 'Gordon Browns speech in full', *BBC News*, 24 September 2007, [internet site]. Available: <http://news.bbc.co.uk/1/hi/uk_politics/7010664.stm> Accessed 8 July 2013.

Brown, Paul (2002). 'New Runway Puts 15,000 Homes at Risk', *The Guardian*, 21 September 2002, [internet site]. Available: <http://www.guardian.co.uk/uk/2002/sep/21/transport.world1> Accessed 28 April 2012.

Burke, Ted (2006). 'Big Ask Public Meeting Lewisham Friends of the Earth', [internet site]. Available: <http://www.foe.co.uk/resource/reports_on_events/lewisham_toptips.pdf> Accessed 22 December 2012.

Burke, Tom (2010). 'Without a True Green Investment Bank There Can Be No Environmental Progress', *The Guardian*. 7 October 2010, [internet site]. Available: <http://www.guardian.co.uk/environment/cif-green/2010/oct/07/green-investment-bank-environment> Accessed 16 August 2012.

Burstein, Paul (1991). 'Policy Domains: Organization, Culture, and Policy Outcomes', *Annual Review of Sociology*, 17: 327–50.

Burstein, Paul (1998) 'Bringing the Public Back In: Should Sociologists Consider the Impact of Public Opinion on Public Policy?', *Social* Forces, 77(1): 27–62.

Burstein, Paul (1999). 'Social Movements and Public Policy', in Marco Giugni, Doug McAdam, and Charles Tilly (eds.) *How Social Movements Matter*, Minneapolis: University of Minnesota Press, pp. 3–21.

Burstein, Paul (2003). 'The Impact of Public Opinion on Public Policy: A Review and an Agenda', *Political Research Quarterly*, 56: 29–40.

Burstein, Paul and April Linton (2002). 'The Impact of Political Parties, Interest Groups, and Social Movement Organizations on Public Policy: Some Recent Evidence and Theoretical Concerns', *Social Forces*, 81(2): 381–408.

Busby, Joshua (2010). *Moral Movements and Foreign Policy*, Cambridge: Cambridge University Press.

Cairns, Sally and Carey Newson (2006). *Predict and Decide: Aviation, Climate Change and UK Policy*, Oxford: Environmental Change Institute.

Cameron, David (2005a). 'Speech at 2005 Conservative Conference in Blackpool', *Total Politics* [internet site]. Available: <http://www.totalpolitics.com/speeches/conservative/conservative-party-conference-general-speeches/34983/speech-at-2005-conservative-conference-in-blackpool.thtml>Accessed 16 August 2014.

Cameron, David (2005b). 'David Cameron: Change Our Political System and Our Lifestyles', *The Independent* [internet site]. Available: <http://www.independent.co.uk/voices/commentators/david-cameron-change-our-political-system-and-our-lifestyles-513423.html> Accessed 16 August 2014.

Cameron, David (2006). 'The Planet First, Politics Second', *The Independent on Sunday*, 3 September.

Čapek, Stella M. (1993). 'The "Environmental Justice" Frame: A Conceptual Discussion and an Application', *Social Problems*, 40(1): 5–24.

Carter, Neil (2006). 'Party Politicization of the Environment in Britain', *Party Politics*, 12(6): 747–67.

Carter, Neil (2008). 'Combating Climate Change in the UK: Challenges and Obstacles', *The Political Quarterly*, 79(2): 194–205.

Carter, Neil and Michael Jacobs (2013). 'Explaining Radical Policy Change: The Case of Climate Change and Energy Policy Under the British Labour Government 2006–10', *Public Administration*, doi: 10.1111/padm.12046.

Carvalho, Anabela (2000). 'Environmental Organizations and the Discursive Construction of Climate Change. Re-reading Activism in the British Press', paper presented at the 41st convention of the International Studies Association, Los Angeles, 15–18 March.

Cass, Loren R. (2006). *The Failures of American and European Climate Policy: International Norms, Domestic Politics, and Unachievable Commitments*, Albany: State University of New York Press.

Chasek, Pamela, David Leonard Downie, Kevin Baumert, Sean Clark, Joshua Tosteson, Les Bissell, Johanna Hjerthen, Johanna Hjerthen, Balachandar Jayaraman, Elizabeth Karkus, John Leahy, and Gerald Mulder (1998). 'European Union Views on International Greenhouse Gas Emissions Trading', Columbia University School of International and Public Affairs Environmental Policy Studies Working Paper 3.

chiswickw4.com (2002). 'Hundreds Turn Out for Anti-Third Runway Protest', [internet site]. Available: <http://www.chiswickw4.com/default.asp?section=info&page=evhacan1.htm> Accessed 5 July 2012.

Christianson, Gale E. (1999). *Greenhouse: The 200-Year Story of Global Warming*, London: Constable.

Clark, Andrew (2004). 'New Doubt Cast on Heathrow Expansion', *The Guardian*, 19 February 2004, [internet site]. Available: <http://www.guardian.co.uk/business/2004/feb/19/theairlineindustry.greenpolitics> Accessed 28 April 2012.

Clasper, Mike (2006). 'Air Passengers Must Cover the Cost of Carbon Emissions', *The Guardian*, 5 April 2006, [internet site]. Available: <http://www.guardian.co.uk/society/2006/apr/05/guardiansocietysupplement9> Accessed 28 April 2012.

Clegg, Nick (2011). 'Deputy Prime Ministers Speech on Green Growth at the Climate Change Capital', [internet site]. Available: <http://www.dpm.cabinetoffice.gov.uk/news/deputy-prime-minister-s-speech-green-growth-climate-change-capital> Accessed 16 August 2012.

Client Earth (2011). 'Towards the Green Investment Bank Act: Legislation to Secure the Mandate and Governance of the UK's Green Investment Bank', [internet site]. Available: <http://www.transformuk.org/attachments/products/33/6clientearth-report-towards-a-green-investment-bank-act.pdf> Accessed 7 September 2012.

Client Earth and Transform UK (2012). 'Summary Critique of the Legislation Establishing the Green Investment Bank', [internet site]. Available: <http://www.clientearth.org/reports/critique-of-the-legislation-establishing-the-gib-final.pdf> Accessed 5 September 2012.

Climate Action Network (2008). 'People Live Here, You Know', *ECO*, CXVI(5): 2.

Climate Action Network (2009a). '575,000 Americans Demand Stronger Action', *ECO*, CXXII(10): 3.

Climate Action Network (2009b). 'Outrage over Lockout', *ECO*, CXXII(11): 4.

Climate Action Network (2009c). 'Oh No, Not Nukes Again!', *ECO*, CXXII(11): 2.

Climate Action Network Canada (2008a). 'Poll: Canadians Want Action on Global Warming Despite Economic Downturn', news release, [internet site]. Available: <http://climateactionnetwork.ca/archive/e/news/2008/cc-poll-2008-12-02.html> Accessed 29 June 2014.

Climate Action Network Canada (2008b). 'Canadian Arctic Indigenous Peoples and Environmental Groups Launch United Call for Action on Climate Change at United Nations Negotiations', news release, [internet site]. Available: < http://climateactionnetwork.ca/archive/e/news/2008/cop14-groups-arctic-2008-12-05.html> Accessed 29 May 2014.

Climate Action Network Canada (2008c). 'Statement by Canadian Environmental Organizations at the U.N. Climate Talks in Poznan', news release, [internet site]. Available: <http://climateactionnetwork.ca/archive/e/news/2008/cop14-statement-2008-12-11.html> Accessed 29 May 2014.

Climate Change Bill (2005). London: HMSO.

Climate Justice Now! (2007). 'CJN! Founding Press Release', [internet site]. Available: <http://www.climate-justice-now.org/cjn-founding-press-release> Accessed 29 May 2014.

Climate Justice Now! (2009). 'CJN! Final Statement in Copenhagen', [internet site]. Available: <http://www.climate-justice-now.org/cjn-final-statement-in-copenhagen/> Accessed 23 August 2014.

Climate Radio (2007). '#4: Climate Bill, Low Carbon Show', [internet site]. Available: <http://climateradio.org/low-carbon-show-4-climate-bill> Accessed 29 May 2013.

Conservative Party (2005). 'Conservative Election Manifesto 2005', [internet site]. Available: <http://newsimg.bbc.co.uk/1/shared/bsp/hi/pdfs/CON_uk_manifesto.pdf> Accessed 29 August 2013.

Corell, Elisabeth and Michele M. Betsill (2001). 'A Comparative Look at NGO Influence in International Environmental Negotiations: Desertification and Climate Change', *Global Environmental Politics*, 1(4): 86–107.
Coumou, Dim and Stefan Rahmstorf (2012). 'A Decade of Weather Extremes', *Nature Climate Change*, 2(7): 491–96.
Darling, Alistair (2011). *Back from the Brink: 1,000 Days at Number 11*, London: Atlantic Books.
Davenport, Deborah (2008). 'The International Dimension of Climate Policy', in Hugh Compston and Ian Bailey (eds.) *Turning Down the Heat: The Politics of Climate Policy in Affluent Democracies*, Basingstoke: Palgrave Macmillan, pp. 48–62.
DECC (2012). *International Aviation and Shipping Emissions and the UK's Carbon Budgets and 2050 Target*, 14D/477, London: Crown Copyright.
Declaration: Tokyo Summit Conference (1979). G7: Tokyo Summit Communique, 28–29 June.
Depledge, Joanna (2005). *The Organization of Global Negotiations: Constructing the Climate Regime*, London: Earthscan.
Dessler, Andew E. and Edward A. Parson (2010). *The Science and Politics of Global Climate Change: A Guide to the Debate*, Cambridge: Cambridge University Press.
DfT (2003). *The Future of Air Transport*, London: HMSO.
DfT (2008). *Adding Capacity at Heathrow Airport: Report on Consultation Responses*, London: Detica Group Plc.
Diani, Mario (1992). 'The Concept of Social Movement', *The Sociological Review*, 40(1): 1–25.
Doherty, Brian (2000). 'Manufactured Vulnerability: Protest Camp Tactics', in Benjamin Seel, Matthew Patterson and Brian Doherty (eds.) *Direct Action in British Environmentalism*, London: Routledge.
Doherty, Sharon (2008). *Heathrow's Terminal 5: History in the Making*, Chicester: John Wiley & Sons, Ltd.
Done, Kevin and Roger Blitz (2006). 'Air Passenger Duty Is Doubled', *Financial Times*, 6 December 2006, [internet site]. Available: <http://www.ft.com/cms/s/0/fbbd599c-8557-11db-b12c-0000779e2340.html#axzz1s1Ce1voV> Accessed 28 April 2012.
Dowdeswell, Elizabeth and Richard J. Kinley (1994). 'Constructive Damage to the Status Quo', in Irving M. Mintzer and J. Amber Leonard (eds.) *Negotiating Climate Change: The Inside Story of the Rio Convention*, Cambridge: Cambridge University Press, pp. 113–28.
Downey, Dennis J. and Deana A. Rohlinger (2008). 'Linking Strategic Choice with Macro-Organizational Dynamics: Strategy and Social Movements Articulation', *Research in Social Movements, Conflicts and Change*, 28: 3–38.
Earl, Jennifer (2006). 'The Cultural Consequences of Social Movements', in David A. Snow, Sarah A. Soule, and Hanspeter Kriesi (eds.) *The Blackwell Companion to Social Movements*, Malden: Blackwell, pp. 508–30.
Earth Negotiations Bulletin (no date a) 'Adequacy of Commitments', [internet site]. Available: <http://www.iisd.ca/vol12/1221013e.html> Accessed 10 September 2014.
Earth Negotiations Bulletin (no date b) '*Ad Hoc* Group on the Berlin Mandate', [internet site]. Available: <http://www.iisd.ca/vol12/1232004e.html> Accessed 10 September 2014.

Environmental Audit Committee (2011). *Second Report – The Green Investment Bank,* [internet site]. Available: <http://www.publications.parliament.uk/pa/cm201011/cmselect/cmenvaud/505/50502.htm> Accessed 16 August 2012.

Enterprise and Regulatory Reform Bill (2013). *Marshalled List of Amendments to Be Moved on Report,* [internet site]. Available: <http://www.publications.parliament.uk/pa/bills/lbill/2012-2013/0083/amend/ml083-i.htm> Accessed 12 July 2013.

Eisinger, Peter K. (1973). 'The Conditions of Protest Behavior in American Cities', *The American Political Science Review*, 67(1): 11–28.

Ernst & Young (2010). *Capitalising the Green Investment Bank: Key Issues and Next Steps,* [internet site]. Available: <http://www.green-alliance.org.uk/uploadedFiles/Themes/Sustainable_Economy/Capitalising%20the%20Green%20Investment%20Bank.pdf> Accessed 5 September 2012.

European Commission (2014). 'Illustrating Public Concern about Climate Change', [internet site]. Available: <http://ec.europa.eu/environment/archives/funding/projects/2000/6.htm> Accessed 5 September 2014.

Event Magazine (2008). 'Showcase: The Big Ask – Gig Focuses Attention on Carbon Emissions', [internet site]. Available: <http://www.eventmagazine.co.uk/news/features/showcase/828603/Showcase-big-ask---Gig-focuses-attention-carbon-emissions> Accessed 8 February 2012.

Falkner, Robert, John Vogler, and Hannes Stephan (2011). 'International Climate Policy after Copenhagen: Toward a "Building Blocks" Approach' in David Held, Angus Hervey, and Marika Theros (eds.) *The Governance of Climate Change: Science, Economics, Politics & Ethics*, Cambridge: Polity Press.

Faulkner, Hugh (1994). 'Some Comments on the INC Process', in Irving M. Mintzer and J. Amber Leonard (eds.) *Negotiating Climate Change: The Inside Story of the Rio Convention*, Cambridge: Cambridge University Press, pp. 229–38.

Featherstone, Lynne (2003). 'Climate Change and the Big Ask', [internet site]. Available: <http://www.lynnefeatherstone.org/2006/10/climate-change-and-big-ask.htm> 12 December 2012.

Ferguson, Thomas, Christian Genest, and Marc Hallin (2011). 'Kendall's tau for autocorrelation', *The Canadian Journal of Statistics*. 65, [internet site]. Available: <http://escholarship.org/uc/item/3p91609d>.

Fisher, Dana R. (2010). 'COP-15 in Copenhagen: How the Merging of Movements Left Civil Society Out in the Cold', *Global Environmental Politics,* 10(2): 11–17.

Fitzmaurice, Malgosia and Olufemi Elias (2005). *Contemporary Issues in the Law of Treaties*, Utrecht: Eleven International Publishing.

Foreman, Esther (2012). *Peering In: An Analysis of Public and Charity Sector Lobbying in the House of Lords,* [internet site]. Available: <http://www.cloresocialleadership.org.uk/userfiles/Peering_In_FINAL_SR.pdf> Accessed 20 April 2013.

Friends of the Earth (2005). 'Yorke Backs Climate Change Campaign', [internet site]. Available: <http://www.foe.co.uk/northern_ireland/press_releases/yorke_backs_climate.html> Accessed 8 February 2012.

Friends of the Earth (2006a). 'New Dates Announced for "The Big Ask Live"', [internet site]. Available: <http://www.foe.co.uk/resource/press_releases/new_dates_announced_for_th_14032006.html> Accessed 23 December 2012.

Friends of the Earth (2006b). 'Lobbying Your MP', [internet site]. Available: <http://www.foe.co.uk/news/lobbying_mp.html> Accessed 23 December 2012.

Friends of the Earth (2006c). 'The Big Month Hits Halfway', [internet site]. Available: <http://www.foe.co.uk/news/bigmonth_halfway.html> Accessed 23 December 2012.

Friends of the Earth (2006d). 'UK Carbon Emissions Still Rising', [internet site]. Available: <http://www.foe.co.uk/resource/press_releases/uk_carbon_emissions_still_20102006.html> Accessed 23 December 2012.

Friends of the Earth (2006e). 'The Big Ask Climate Debate Blair', [internet site]. Available: http://www.foe.co.uk/news/blair_juniper_tony_blair.html Accessed 23 December 2012.

Friends of the Earth (2006f). 'Friends of the Earth Secures Climate Change Bill', [internet site]. Available: <http://www.foe.co.uk/news/gov_climate_bill.html> Accessed 23 December 2012.

Friends of the Earth (2007a). 'As Climate Bill Consultation Ends – Summer Begins', [internet site]. Available: <http://www.foe.co.uk/news/end_consultation.html> Accessed 8 February 2012.

Friends of the Earth (2007b). 'The Big Ask – Big Autumn Push', [internet site]. Available: <http://www.foe.co.uk/news/big_push.html> Accessed 8 February 2012.

Friends of the Earth (2008a). 'History of The Big Ask', [internet site]. Available: <http://www.foe.co.uk/news/big_ask_history_15798.html> Accessed 8 February 2012.

Friends of the Earth (2008b). 'Climate Change and Energy Bills Get Royal Assent', [internet site]. Available: <http://www.foe.co.uk/resource/press_releases/royal_assent_26112008.html> Accessed 8 February 2012.

Friends of the Earth (2009). 'Heathrow Terminal 5 and Runway 3', [internet site]. Available: <http://www.foe.co.uk/resource/media_briefing/heathrow_broken_promises.pdf> Accessed 11 May 2012.

Friends of the Earth (2010a). 'Liberal Democrats Have Greenest Manifesto of Main Parties', [internet site]. Available: <http://www.foe.co.uk/resource/press_releases/manifesto_assessment_election2010_26042010.html> Accessed 12 July 2013.

Friends of the Earth (2010b). 'Councils Key to Meeting UK's Green Energy Target, Report Warns', [internet site]. Available: <http://www.foe.co.uk/resource/press_releases/renewables_energy_target_27072010> Accessed 12 July 2014.

Friends of the Earth (2010c). 'Cautious Welcome for Climate Plans', [internet site]. Available: <http://www.foe.co.uk/news/24787> Accessed 12 July 2014.

Friends of the Earth (2011a). 'Charities Question Cameron's "Greenest Government Ever" Claim One Year On' [internet site]. Available: <http://www.foe.co.uk/resource/press_releases/greenest_govt_ever_coalition_14052011.html> Accessed 5 September 2012.

Friends of the Earth (2011b). 'Cameron's Council Calls for Stronger Energy Bill', [internet site]. Available: <http://www.foe.co.uk/news/witney_meeting_31183.html> Accessed 17 April 2013.

Friends of the Earth (2011c). 'Budget Fails to Tackle Oil Dependency', [internet site]. Available: <http://www.foe.co.uk/news/budget_news_reaction_30099.html> Accessed 17 April 2013.

Friends of the Earth (2012). 'Green Investment Bank HQ to Be in Edinburgh', [internet site]. Available: <http://www.foe.co.uk/resource/press_releases/green_investment_bank_hq_edinbugh_08032012.html> Accessed 5 September 2012.

Friends of the Earth Birmingham (2005). 'MPs Get the Bill for Climate Change', [internet site]. Available: <http://www.birminghamfoe.org.uk/newslet/news0805/STORY_5.HTM> Accessed 22 December 2012.

Friends of the Earth Ealing (2006). 'Climate Change Campaign – Big Ask, Big Month, Big Lobby', [internet site]. Available: <http://www.ealingfoe.org.uk/Newsletters/0607BigAskBigMonthBigLobby.html> Accessed 22 December 2012.

Friends of the Earth Hammersmith and Fulham (2007). 'Local Climate Campaign', [internet site]. Available: <http://www.paulaandsteve.btinternet.co.uk/climatenews2.html> Accessed 29 March 2012.

Friends of the Earth Trust Limited (2008). 'Reports and Accounts', [internet site]. Available: <http://www.charitycommission.gov.uk/Accounts/Ends81%5C0000281681_ac_20080531_e_c.pdf> Accessed 17 December 2012.

Gamson, William A. (1975). *The Strategy of Social Protest*, Dorsey Press: Homewood.

Gamson, William A. and David S. Meyer (1999). 'Framing Political Opportunities', in Doug McAdam, John D. McCarthy, and Mayer N. Zald (eds.) *Comparative Perspectives on Social Movements: Political Opportunities, Mobilizing Structures, and Cultural Framings*, Cambridge: Cambridge University Press, pp. 275–90.

Ganz, Marshall (2000). 'Resources and Resourcefulness: Strategic Capacity in the Unionization of California Agriculture, 1959–1966', *American Journal of Sociology*, 105(4): 1003–62.

Ganz, Marshall (2004). 'Why David Sometimes Wins: Strategic Capacity in Social Movements', in Jeff Goodwin and James M. Jasper (eds.) *Rethinking Social Movements: Structure, Meaning and Emotion*, Lanham, MD: Rowman & Littlefield Publishers, pp. 177–98.

Giugni, Marco G. (1998). 'Was It Worth the Effort? The Outcomes and Consequences of Social Movements', *Annual Review of Sociology*, 24: 371–93.

Giugni, Marco (1999). 'Introduction – How Social Movements Matter: Past Research, Present Problems, Future Developments', in Marco Giugni, Doug McAdam, and Charles Tilly (eds.) *How Social Movements Matter*, Minneapolis: University of Minnesota Press, pp. xiii–xxxiii.

Giugni, Marco (2004). *Social Protest and Policy Change: Ecology, Antinuclear, and Peace Movements in Comparative Perspective*, Lanham, MD: Rowman & Littlefield.

Giugni, Marco G. (2006). 'Personal and Biographical Consequences', in David A. Snow, Sarah. A. Soule, and Hanspeter Kriesi (eds.) *The Blackwell Companion to Social Movements*, Malden: Blackwell, pp. 489–507.

Giugni, Marco and Sakura Yamasaki (2009). 'The Policy Impact of Social Movements: A Replication Through Qualitative Comparative Analysis', *Mobilization: An International Journal*, 14(4): 467–84.

Goodwin, Jeff and James M. Jasper (2004). 'Caught in a Winding, Snarling Vine: The Structural Bias of Political Process Theory', in Jeff Goodwin and James M. Jasper (eds.) *Rethinking Social Movements: Structure, Meaning and Emotion*, Lanham: Rowman & Littlefield Publishers.

Gore, Georgiana (2010). 'Flash Mob Dance and the Territorialisation of Urban Movement', *Anthropological Notebooks*, 16(3): 125–31.

Green Alliance (2009). 'When Will the Low-Carbon Bank Open for Business?', [internet site]. Available: <http://www.green-alliance.org.uk/grea1.aspx?id=4629> Accessed 5 September 2012.

Green Alliance (2010). *The Comprehensive Spending Review: Analysis of Governments' Low Carbon Spending,* [internet site]. Available: <http://www.green-alliance.org.uk/grea1.aspx?id=5218> Accessed 5 September 2012.

Green Investment Bank Commission (2010). *Unlocking Investment to Deliver Britain's Low Carbon Future,* [internet site]. Available: <http://www.climatechange-capital.com/images/docs/publications/unlocking-investment.pdf> Accessed 25 September 2013.

Greenpeace (no date). 'Contaminated? There's a Toxic Plot in the Conservative Party', [internet site]. Available: <http://www.greenpeace.org.uk/energygate> Accessed 25 September 2013.

Greenpeace (2007a). 'Greenpeace Guide to Kyoto, Bali, APEC, the G8 and Major Emitters meeting', [internet site]. Available: <http://www.greenpeace.org/international/Global/international/planet-2/report/2007/10/guide-to-major-climate-meetings.pdf> Accessed 28 August 2014.

Greenpeace (2007b). 'Greenpeace Gives Away Free Train Tickets at Airports across the UK', [internet site]. Available: <http://www.greenpeace.org.uk/blog/climate/greenpeace-sets-up-climate-ticket-exchanges-across-the-uks-airports-20070619< Accessed 11 May 2012.

Greenpeace (2007c). 'BAA Files Reveal Collusion with Government over Heathrow 3rd Runway Plans', [internet site]. Available: <http://www.greenpeace.org.uk/media/reports/baa-files> Accessed 28 April 2012.

Greenpeace (2009). 'A Risky Business: Government Spin Plan over Heathrow Revealed', [internet site]. Available: <http://www.greenpeace.org.uk/blog/climate/risky-business-20090327> Accessed 07 May 2012.

Greenpeace (2010a). 'Green Campaigners Scale Treasury Ahead of Government Spending Review', [internet site]. Available: <http://www.greenpeace.org.uk/media/press-releases/green-campaigners-scale-treasury-ahead-government-spending-review-20101019> Accessed 5 September 2012.

Greenpeace (2010b). 'Greenpeace Response to Comprehensive Spending Review', [internet site]. Available: <http://www.greenpeace.org.uk/media/press-releases/greenpeace-response-comprehensive-spending-review-20101020> Accessed 5 September 2012.

Greenpeace (2010c). 'The One Bank We Really Should Be Saving', [internet site]. Available: <http://www.greenpeace.org.uk/blog/climate/one-bank-we-really-should-be-saving-20101217> Accessed 5 September 2012.

The Guardian (2010). 'I Want a Green Bank as Soon as Possible', [internet site]. Available: <http://www.guardian.co.uk/environment/2010/dec/17/i-want-green-bank-soon> Accessed 16 August 2012.

Gulbrandsen, Lars H. and Steinar Andresen (2004). 'NGO Influence in the Implementation of the Kyoto Protocol: Compliance, Flexibility Mechanisms, and Sinks', *Global Environmental Politics,* 4(4): 54–75.

HACAN ClearSkies (2002a). 'Government Should Come Clean on Its Intentions', [internet site]. Available: <http://web.archive.org/web/20070710230545/http://www.hacan.org.uk/news/press_releases.php?id=45> Accessed 11 August 2014.

HACAN ClearSkies (2002b). '3rd Runway Home Demolitions', [internet site]. Available: <http://www.hacan.org.uk/news/press_releases.php?id=13> Accessed 11 May 2012.

HACAN ClearSkies (2005a). 'Judicial Review Judgment on Aviation White Paper "Setback" for Government', [internet site]. Available: <http://www.hacan.org.uk/news/press_releases.php?id=111> Accessed 11 May 2012.

HACAN ClearSkies (2005b). 'Appointment of BA Chief to Advise Ministers "Cements Links between the Aviation Industry and Government"', [internet site]. Available: <http://www.hacan.org.uk/news/press_releases.php?id=112> Accessed 11 May 2012.

HACAN ClearSkies (2005c). 'Angry Residents Give Rod Eddington a Taste of His Own Medicine', [internet site]. Available: <http://www.hacan.org.uk/news/press_releases.php?id=121> Accessed 11 May 2012.

HACAN ClearSkies (2009). 'Government Majority Slashed by Two Thirds on Heathrow', [internet site]. Available: <http://www.hacan.org.uk/news/press_releases.php?id=240> Accessed 11 May 2012.

Hall, Nina L. and Ros Taplin (2007). 'Solar Festivals and Climate Bills: Comparing NGO Climate Change Campaigns in the UK and Australia', *Global Warming: Energy Security or Energy Sovereignty?* University of Sydney, 2 March 2007.

Harper, Keith (2001). 'Byers Keeps Third Runway Option Open', *The Guardian*, 21 November 2001, [internet site]. Available: <http://www.guardian.co.uk/uk/2001/nov/21/transport.publicservices2> Accessed 27 April 2012.

Harper, Keith and Peter Hetherington (2002). 'Inspector Admits Third Heathrow Runway Possible in New Circumstances', *The Guardian*, 24 January 2002, [internet site]. Available: <http://www.guardian.co.uk/politics/2002/jan/24/publicservices.uk> Accessed 27 April 2012.

Harrison, Carolyn M., Jacquelin Burgess, and Petra Filius (1996). 'Rationalizing Environmental Responsibilities: A Comparison of Lay Publics in the UK and the Netherlands', *Global Environmental Change*, 6(3): 215–34.

Harvey, Fiona (2011). 'Budget 2011: Osborne's Green Bank Attacked from All Sides', *The Guardian*, 23 March 2011, [internet site]. Available: <http://www.guardian.co.uk/environment/2011/mar/23/budget-2011-george-osborne-green-bank> Accessed 3 March 2013.

Harvey, Fiona and Damian Carrington (2011). 'Green ISA Plans under Threat', *The Guardian*, 25 February 2011, [internet site]. Available: <http://www.guardian.co.uk/environment/2011/feb/25/green-isa-under-threat> Accessed 3 March 2013.

Hathaway, Will and David S. Meyer (1993–1994). 'Competition and Cooperation in Social Movement Coalitions: Lobbying for Peace in the 1980s', *Berkeley Journal of Sociology*, 38: 157–83.

Hedström, Peter and Petri Ylikoski (2010). 'Causal Mechanisms in the Social Sciences', *Annual Review of Sociology*, 36: 49–67.

Hencke, David (2009). 'Transport Minister Backs £4.5bn Rail Hub for Heathrow', *The Guardian*, 5 January 2009 [internet site]. Available: <http://www.guardian.co.uk/politics/2009/jan/05/heathrow-international-rail-exchange> Accessed 28 April 2012.

Helsel, D. R. and R. M. Hirsch (1992). *Statistical Methods in Water Resources*, Elsevier: Amsterdam.

Henry, Laura A. and Lisa McIntosh Sundstrom (2007). 'Russia and the Kyoto Protocol: Seeking an Alignment of Interests and Image', *Global Environmental Politics*, 7(4): 47–69.

Hewitt, Chris (2012). 'Government and Opposition Back a Green Investment Bank', Green Alliance [internet site]. Available: <http://www.green-alliance.org.uk/grea1.aspx?id=4771> Accessed 5 September 2012.

HM Treasury (2005). *Britain Meeting the Global Challenge: Enterprise, Fairness and Responsibility, Pre-Budget Report*, London: HMSO.

HM Treasury (2010). *Spending Review 2010*, Cm 7942, London: Crown Copyright.

Hojnacki, Marie (1998). 'Organized Interests Advocacy Behavior in Alliances', *Political Research Quarterly*, 51(2): 437–59.

Holmes, Ingrid and Nick Mabey (2009). *Accelerating Green Infrastructure Financing: Outline Proposals for UK Green Bonds and Infrastructure Bank*, [internet site]. Available: <http://www.climatechangecapital.com/images/docs/publications/4.pdf> Accessed 15 September 2012.

Holmes, Ingrid and Nick Mabey (2010). *Accelerating the Transition to a Low Carbon Economy: The Case for a Green Infrastructure Bank*, [internet site]. Available: <http://www.transformuk.org/attachments/products/19/acceleratingthetransitiontoalowcarboneconomythecaseforagreeninfrastructurebank.pdf> Accessed 5 September 2012.

House of Commons (2006). 'Engagements', vol. 450, col. 299. 11 October 2006, [internet site]. Available: <http://www.publications.parliament.uk/pa/cm200506/cmhansrd/vo061011/debtext/61011-0004.htm> Accessed 21 September 2013.

House of Commons Hansard Parliamentary Debates (2008a). *Climate Change Bill*, vol. 477, col. 37-133.

House of Commons Hansard Parliamentary Debates (2008b). *Climate Change Bill [Lords]*, Clause 2, 26 June 2008.

House of Commons Hansard Parliamentary Debates (2012). *Enterprise and Regulatory Reform Bill*, vol. 551, col. 327-445.

House of Lords (2010). *Airports: New Runways*, vol. 718, col. 1285-1288.

House of Lords Hansard Parliamentary Debates (2012) *Enterprise and Regulatory Reform Bill*, vol. 740, col. 1517-1612.

Hulme, Mike (2013). *Exploring Climate Change through Science and in Society: An Anthology of Mike Hulme's Essays, Interviews and Speeches*, Abingdon: Routledge.

INC (1991). *Report of the Intergovernmental Negotiating Committee for a Framework Convention on Climate Change on the Work of its First Session*, Document A/AC.237/6, [internet site]. Available: <http://unfccc.int/resource/docs/a/06.pdf> Accessed 10 August 2014.

IPCC (2014). *Climate Change 2014: Impacts, Adaptation, and Vulnerability. Part A: Global and Sectoral Aspects*. Contribution of Working Group II to the Fifth Assessment Report of the Intergovernmental Panel on Climate Change, Cambridge: Cambridge University Press.

Ipsos MORI (1998). 'Citizens Want Tough Environmental Action Now', [internet site]. Available: <http://www.ipsos-mori.com/researchpublications/researcharchive/2017/Citizens-Want-Tough-Environmental-Action-Now.aspx> Accessed 29 March 2012.

Ipsos MORI (2002). 'Public Uncertainty over Environmental Issues', [internet site]. Available: <http://www.ipsos-mori.com/researchpublications/researcharchive/1072/Public-Uncertainty-over-Environmental-Issues.aspx> Accessed 29 March 2012.

Ipsos MORI (2004). 'Saving the World Will Have To Wait – Most Americans Need Convincing', [internet site]. Available: <http://www.ipsos-mori.com/researchpublications/researcharchive/609/Saving-The-World-Will-Have-To-Wait-Most-Americans-Need-Convincing.aspx> Accessed 29 March 2012.

Ipsos MORI (2005a). 'MORI Political Monitor January 2005', [internet site]. Available: <http://www.ipsos-mori.com/researchpublications/researcharchive/507/MORI-Political-Monitor-January-2005.aspx> Accessed 29 March 2012.

Ipsos MORI (2005b). 'MORI Political Monitor, February 2005', [internet site]. Available: <http://www.ipsos-mori.com/researchpublications/researcharchive/508/MORI-Political-Monitor-February-2005.aspx> Accessed 29 March 2012.

Ipsos MORI (2005c). 'MORI Political Monitor March', [internet site]. Available: <http://www.ipsos-mori.com/researchpublications/researcharchive/509/MORI-Political-Monitor-March.aspx> Accessed 29 March 2012.

Ipsos MORI (2005d). 'MORI Political Monitor May', [internet site]. Available: <http://www.ipsos-mori.com/researchpublications/researcharchive/510/MORI-Political-Monitor-May.aspx> Accessed 29 March 2012.

Ipsos MORI (2005e). 'MORI Political Monitor June 2005', [internet site]. Available: <http://www.ipsos-mori.com/researchpublications/researcharchive/511/MORI-Political-Monitor-June-2005.aspx> Accessed 29 March 2012.

Ipsos MORI (2005f). 'MORI Political Monitor July 2005', [internet site]. Available: <http://www.ipsos-mori.com/researchpublications/researcharchive/512/MORI-Political-Monitor-July-2005.aspx> Accessed 29 March 2012.

Ipsos MORI (2005g). 'MORI Political Monitor August 2005', [internet site]. Available: <http://www.ipsos-mori.com/researchpublications/researcharchive/513/MORI-Political-Monitor-August-2005.aspx> Accessed 29 March 2012.

Ipsos MORI (2005h). 'MORI Political Monitor September 2005', [internet site]. Available: <http://www.ipsos-mori.com/researchpublications/researcharchive/515/MORI-Political-Monitor-September-2005.aspx> Accessed 29 March 2012.

Ipsos MORI (2005i). 'MORI Political Monitor October 2005', [internet site]. Available: <http://www.ipsos-mori.com/researchpublications/researcharchive/516/MORI-Political-Monitor-October-2005.aspx> Accessed 29 March 2012.

Ipsos MORI (2005j). 'Ipsos MORI Political Monitor November 2005', [internet site]. Available: <http://www.ipsos-mori.com/researchpublications/researcharchive/517/Ipsos-MORI-Political-Monitor-November.aspx> Accessed 29 March 2012.

Ipsos MORI (2007). 'Political Commentary – Public: Government Should Intervene on Climate Change … Just Don't Tax Us', [internet site]. Available: <http://www.ipsos-mori.com/newsevents/ca/277/Political-Commentary-Public-Government-Should-Intervene-On-Climate-Change-8230-Just-Dont-Tax-Us.aspx> Accessed 29 March 2012.

Ipsos MORI (2009). 'Economy vs. Environment – A Global Tension', [internet site]. Available: <http://www.ipsos-mori.com/researchspecialisms/socialresearch/specareas/environment/understandingenvironmentnewsletter/economyvsenvironment.aspx> Accessed 29 March 2012.

Jasper, James M. (2004). 'A Strategic Approach to Collective Action: Looking for Agency In Social-Movement Choices', *Mobilization*, 9(1): 1–16.

Johnston, Ron, David Rossiter, and Charles Pattie (2006). 'Disproportionality and Bias in the Results of the 2005 General Election in Great Britain: Evaluating the Electoral Systems Impact', *Journal of Elections, Public Opinion and Parties*, 16, 37–54.

Joint Committee on the Draft Climate Change Bill (2007a). *Draft Climate Change Bill: Report of Session 2006–07, Vol II, HL Paper 170-II, HC 542-II*: London: TSO.

Joint Committee on the Draft Climate Change Bill (2007b). *Draft Climate Change Bill, Vol I, HL Paper 170-I, HC 542-I*: London: TSO.

Jones, Robin Russell (1985). 'The Murky Politics that Keep Britain the Dirty Man of Europe', *The Guardian*, 22 November, p. 18.

Jowit, Juliette and Gaby Hinsliff (2006). 'Fury Greets New Plan for Heathrow Expansion', *The Observer*, 1 January, [internet site]. Available: <http://www.guardian.co.uk/business/2006/jan/01/theairlineindustry.politics> Accessed 28 April 2012.

Juniper, Tony (2008). 'Friends of the Earth Director Tony Juniper Sensationally Quits – and Attacks Hypocritical "Green" Celebrities', *Daily Mail*, 27 January [internet site]. Available: <http://www.guardian.co.uk/business/2006/jan/01/theairlineindustry.politics> Accessed 28 April 2012.

Keck, Margaret E. and Kathryn Sikkink (1999). 'Transnational Advocacy Network in International and Regional Politics', *International Social Science Journal*, 51(159): 89–101.

Khor, Martin (2012). *Reaffirming the Environment-Development Nexus of UNCED 1992*, Penang: Third World Network.

Kingdon, John W. (1995). *Agendas, Alternatives, and Public Policies*, New York: HarperCollins.

Kirkup, James (2009). 'Heathrow Airport to Get Third Runway, Geoff Hoon Announces', *Telegraph*, 15 January, [internet site]. Available: <http://www.telegraph.co.uk/travel/travelnews/4246548/Heathrow-airport-to-get-third-runway-Geoff-Hoon-announces.html> Accessed 07 May 2012.

Kitschelt, Herbert P. (1986). 'Political Opportunity Structures and Political Protest: Anti-Nuclear Movements in Four Democracies', *British Journal of Political Science*, 16, 57–85.

Knopf, Jeffrey W. (1998). *Domestic Society and International Cooperation: The Impact of Protest on US Arms Control Policy*, Cambridge: Cambridge University Press.

Kolb, Felix (2007). *Protest and Opportunities: The Political Outcomes of Social Movements*, Frankfurt: Campus Verlag.

Koopmans, Ruud (1993). 'The Dynamics of Protest Waves: West Germany, 1965 to 1989', *American Sociological Review*, 58, 637–58.

Koopmans, Ruud (1999). 'Political. Opportunity. Structure. Some Splitting to Balance the Lumping', *Sociological Forum*, 14(1): 93–105.

Kraft, Michael E. and Norman J. Vig (2006). 'Environmental Policy from the 1970s to the Twenty-First Century', in Norman J. Vig and Michael E. Kraft (eds.) *Environmental Policy: New Directions for the Twenty-First Century*, Washington: CQ Press, pp. 1–33.

Kutney, Gerald (2014). *Carbon Politics and the Failure of the Kyoto Protocol*, Abingdon: Routledge.

Labour Party (2005). 'The Labour Party Manifesto 2005: Britain Forward Not Back', [internet site]. Available: <http://ucrel.lancs.ac.uk/wmatrix/tutorial/labour%20manifesto%202005.pdf> Accessed 17 August 2014.

Lemonick, Michael D. (1989). 'Global Warming Feeling the Heat – The Problem: Greenhouse Gases Could Create a Climatic Calamity', *TIME*, Vol. 133, Issue 1, p. 36–41.

Liberal Democrats (2005). 'The Real Alternative', Party Manifesto [internet site]. Available: <http://ucrel.lancs.ac.uk/wmatrix/tutorial/libdem%20manifesto%202005.pdf> Accessed 17 August 2014.

Liberal Democrats (2010). 'Manifesto 2010', [internet site]. Available: <http://www.politicsresources.net/area/uk/ge10/man/parties/libdem_manifesto_2010.pdf> Accessed 17 August 2014.

Little, Paul E. (1995). 'Ritual, Power and Ethnography at the Rio Earth Summit', *Critique of Anthropology*, 15(3): 265–88.

Lord Turner of Ecchinswell (2008). 'Interim Advice by the Committee on Climate Change, Letter to Secretary of State DECC Ed Miliband', [internet site]. Available: <http://www.theccc.org.uk/wp-content/uploads/2013/03/Interim-report-letter-to-DECC-SofS-071008.pdf> Accessed 17 December 2012.

Luders, Joseph E. (2010). *The Civil Rights Movement and the Logic of Social Change*, Cambridge: Cambridge University Press.

MacDonald, Douglas and Smith, Heather A. (1999–2000). 'Promises Made, Promises Broken – Questioning Canada's Commitments to Climate Change', *International Journal*, 55(1): 107–24.

Maney, Gregory M., Kenneth T. Andrews, Rachel V. Kutz-Flamenboum, Deana A. Rohlinger, and Jeff Goodwin (2012). 'An Introduction to Strategies for Social Change', in Gregory M. Maney, Rachel V. Kutz-Flamenboum, Deana A. Rohlinger, and Jeff Goodwin (eds.) *Strategies for Social Change*, Minneapolis: University of Minnesota Press.

Marston, Paul (2003). 'Airports "Need Three New Runways"', *The Telegraph*, 13 May, [internet site]. Available: <http://www.telegraph.co.uk/news/uknews/1429911/Airports-need-three-new-runways.html> Accessed 7 February 2013.

McAdam, Doug (1989). 'The Biographical Consequences of Activism', *American Sociological Review*, 54: 744–60.

McAdam, Doug (1999). 'Conceptual Origins, Current Problems, Future Directions', in Doug McAdam, John D. McCarthy, and Mayer N. Zald (eds.) *Comparative Perspectives on Social Movements: Political Opportunities, Mobilizing Structures, and Cultural Framings*, Cambridge: Cambridge University Press, pp. 23–40.

McAdam, Doug and Yang Su (2002). 'The War at Home: Antiwar Protests and Congressional Voting, 1965 to 1973', *American Sociological Review*, 67, 696–721.

McCammon, Holly J., Karen E. Campbell, Ellen M. Granberg, and Christine Mowery (2001). 'How Movements Win: Gendered Opportunity Structures and U.S. Women's Suffrage Movements, 1866 to 1919', *American Sociological Review*, 66(1): 49–70.

McCarthy, John D. and Mayer N. Zald (1997). 'Resource Mobilization and Social Movements: A Partial Theory', *American Journal of Sociology*, 82(6): 1212–41.

Mcgranahan, Gordon, Deborah Balk, and Bridget Anderson (2007). 'The Rising Tide: Assessing the Risks of Climate Change and Human Settlements in Low Elevation Coastal Zones', *Environment and Urbanization*, 19(1): 17–37.

McCombs, Maxwell E. and Donald L. Shaw (1972). 'The Agenda-Setting Function of Mass Media', *Public Opinion Quarterly*, 36, 176–87.

Meacher, Michael (2008). 'A Flying Leap', *The Guardian*, 12 November, [internet site]. Available: <http://www.guardian.co.uk/commentisfree/2008/nov/12/heathrow-third-runway> Accessed 28 April 2012.

Meyer, David S. (2004). 'Protest and Political Opportunities', *Annual Review of Sociology*, 30, 125–45.

Meyer, David S. and Suzanne Staggenborg (1996) 'Movements, Countermovements, and the Structure of Political Opportunity', *American Journal of Sociology*, 101(6): 1628–60.

Meyer, John W. and Brian Rowan (1997). 'Institutionalized Organizations: Formal Structure as Myth and Ceremony', *American Journal of Sociology*, 83(2): 340–63.

Midttun, Atle and Dieter Rucht (1994). 'Comparing Policy Outcomes of Conflicts over Nuclear Power: Description and Explanation', in Helena Flam (ed.) *States and Anti-Nuclear Movements,* Edinburgh: Edinburgh University Press, pp. 383–415.

Milmo, Dan (2006a). 'Green Policies Will Hurt Economy, Says BA', *The Guardian*, 14 November, [internet site]. Available: <http://www.guardian.co.uk/business/2006/nov/14/britishairways.theairlineindustry> Accessed 28 April 2012.

Milmo, Dan (2006b). 'Don't Listen to Air Industry Doom-Mongers, Ministers Told', *The Guardian*, 5 December [internet site]. Available: <http://www.guardian.co.uk/business/2006/dec/05/theairlineindustry.politics> Accessed 28 April 2012.

Milmo, Dan (2008a). 'Walsh Attacks Cameron over Airport Expansion', *The Guardian*, 26 June, [internet site]. Available: <http://www.guardian.co.uk/business/2008/jun/26/theairlineindustry.britishairwaysbusiness> Accessed 28 April 2012.

Milmo, Dan (2008b). 'Airline Industry Slams Tory Plans to Scrap Third Runway', *The Guardian*, 29 September, [internet site]. Available: <http://www.guardian.co.uk/business/2008/sep/29/britishairways.baa> Accessed 29 April 2012.

Monbiot, George (2009). 'The Third Runway Is the Final Betrayal', *The Guardian*, 15 January, [internet site]. Available: <http://www.guardian.co.uk/commentisfree/2009/jan/15/heathrow-third-runway-labour> Accessed 07 May 2012.

Moore, Charles (2013). *Margaret Thatcher: The Authorized Biography, Volume One: Not for Turning*, London: Penguin.

Mulholland, Hélène (2008). 'Third Runway Would Drive Coach and Horses through Mayor's Green Plans', *The Guardian*, 16 December, [internet site]. Available: <http://www.guardian.co.uk/politics/2008/dec/16/boris-johnson-heathrow> Accessed 28 April 2012.

Murphy, Joe (2012). '"I Wasn't Made Transport Secretary to Push Through Third Runway at Heathrow...All Options Are on the Table"', *London Evening Standard*, 27 September, [internet site]. Available: <http://www.standard.co.uk/news/politics/i-wasnt-made-transport-secretary-to-push-through-third-runway-at-heathrowall-options-are-on-the-table-8181588.html> Accessed 28 August 2014.

Murray, Leo (2007). 'Diary of a Protest', *Index on Censorship*, 36, 22–39.

Newell, Peter (2000). *Climate for Change: Non-state Actors and the Global Politics of the Greenhouse*, Cambridge: Cambridge University Press.

Nulman, E. (2015). 'Dynamic interactions in contentious episodes: social movements, industry, and political parties in the contention over Heathrow's third runway', *Environmental Politics*, DOI:10.1080/09644016.2015.1014657.

Oberthür, Sebastian and Hermann E. Ott (1999). *The Kyoto Protocol: International Climate Policy for the 21*st *Century*, Berlin: Springer-Verlag.

Page, Benjamin I., Robert Y. Shapiro, and Glenn R. Dempsey (1987). 'What Moves Public Opinion', *The American Political Science Review*, 81, 23–44.

Parliament UK (no date). 'Early Day Motion 178', [internet site]. Available: <http://www.parliament.uk/edm/2005-06/178> Accessed 19 September 2013.

Paterson, Matthew (1996). *Global Warming and Global Politics*, London: Routledge.

Pettenger, Mary E. (2007). *The Social Construction of Climate Change: Power, Knowledge, Norms, Discourses*, Aldershot: Ashgate.

Piven, Frances F. and Richard A. Cloward (1977). *Poor Peoples' Movements: Why They Succeed, How they Fail*, New York: Pantheon Books.

Plane Stupid (2005). 'Climate Activists Disrupt International Aviation Conference', [internet site]. Available: <http://www.planestupid.com/?q=content/climate-activists-disrupt-international-aviation-conference-29th-september-2005> Accessed 11 May 2012.

Plane Stupid (2009). 'Activists Accuse Climate Change Secretary of "Hypocracy"', [internet site]. Available: <http://www.planestupid.com/content/activists-accuse-climate-change-secretary-hypocracy> Accessed 11 May 2012.

Populus (2006). 'Political Attitudes, Fieldwork: March 30th 2006 – April 1st 2006', [internet site]. Available: <http://populuslimited.com/preview/poll.php?poll=52> Accessed 11 May 2014.

Porritt, Jonathon (1989). 'The United Kingdom: The Dirty Man of Europe?', *RSA Journal*, 137(5396): 488.

Powell, Dave (2012). '(Green) Banking on Our Politicians', [internet site]. Available: <http://www.foe.co.uk/blog/green_bank_36029.html> Accessed 5 September 2012.

Rahman, Atiq and Annie Roncerel (1994). 'A View from the Ground Up', in Irving M. Mintzer and J. Amber Leonard (eds.) *Negotiating Climate Change: The Inside Story of the Rio Convention*, Cambridge: Cambridge University Press, pp. 239–73.

Raustiala, Kal (2001). 'Nonstate Actors in the Global Climate Regime', in Urs Luterbacher and Detlef F. Sprinz (eds.) *International Relations and Global Climate Change*, Cambridge, Mass: MIT Press, pp. 95–117.

Reimann, Kim D. (2002). 'Building Networks from the Outside In: Japanese NGOs and the Kyoto Climate Change Conference', Faculty Publications, Paper 6, [internet site]. Available: <http://digitalarchive.gsu.edu/political_science_facpub/6> Accessed 5 September 2012.

Rietig, Katharina (2011). 'Public Pressure versus Lobbying – How Do Environmental NGOs Matter Most in Climate Negotiations?', Centre for Climate Change Economics and Policy, Working Paper No. 79, Grantham Research Institute on Climate Change and the Environment, Working Paper No. 70.

Rootes, Christopher A. (1999). 'Political Opportunity Structures: Promise, Problems and Prospects', *La Lettre de la maison Française dOxford*, 10: 75–97.

Rootes, Christopher (2007a). 'Environmental Movements', in David Snow, Sarah A. Soule, and Hanspeter Kriesi (eds.) *The Blackwell Companion to Social Movements*, Malden: Blackwell, pp. 608–40.

Rootes, Christopher (2007b). 'Acting Locally: The Character, Contexts and Significance of Local Environmental Mobilisations', *Environmental Politics*, 16(5): 722–41.

Rootes, Christopher (2011). 'New Issues, New Forms of Action? Climate Change and Environmental Activism in Britain', in Jan W. van Deth and William Maloney (eds.) *New Participatory Dimensions in Civil Society: Professionalization and Individualized Collective Action*, London: Routledge, pp. 46–68.

Rootes, Christopher (2012). 'Climate Change, Environmental Activism and Community Action in Britain', *Social Alternatives*, 31(1): 24–8.

Rootes, Christopher (2013). 'From Local Conflict to National Issue: When and How Environmental Campaigns Succeed in Transcending the Local', *Environmental Politics*, 22(1): 95–114.

Rosenberg, Gerald N. (1993). *The Hollow Hope: Can Courts Bring About Social Change?* University of Chicago Press: Chicago.

Saunders, Clare (2004). *Collaboration, Competition and Conflict: Social Movement and Interaction Dynamics of London's Environmental Movement*. Ph.D thesis, University of Kent.

Sawer, Patrick (2009). 'Geoff Hoon and Emma Thompson Fall Out over Heathrow's Third Runway', *The Telegraph*, 17 January, [internet site]. Available: <http://www.telegraph.co.uk/travel/travelnews/4276800/Geoff-Hoon-and-Emma-Thompson-fall-out-over-Heathrows-third-runway.html> Accessed 07 May 2012.

Schlanger, Zoë (2014). 'Leaked U.N. Report: Climate Change Impacts Already "Inevitable," May Soon Be "Irreversible"', *Newsweek*, 26 August [internet site]. <Available: http://www.newsweek.com/leaked-un-report-climate-change-impacts-already-inevitable-may-soon-be-irreversible-266860> Accessed 11 May 2012.

SchNEWS (2001). 'BONN VOYAGE!', Issue 314/315, [internet site]. Available: <http://www.schnews.org.uk/archive/news314-5.htm> Accessed 11 August 2014.

SchNEWS (2005). '...And Finally...', [internet site]. Available: <http://www.schnews.org.uk/archive/news499.htm> Accessed 11 May 2012.

Seldon, Anthony and Guy Lodge (2011). *Brown at 10*, Hull: Biteback Publishing.

Shaffer, Martin B. (2000). 'Coalition Work Among Environmental Groups: Who Participates?', in Patrick G. Coy (ed.) *Research in Social Movements, Conflicts and Change* Volume 22, JAI Press: Stamford, pp. 111–26.

Sheffield Friends of the Earth (no date). 'Climate Change Campaign 1997', [internet site]. Available: <http://www.sheffieldfoe.pwp.blueyonder.co.uk/campaigns/climate/climateold.htm> Accessed 17 December 2012.

Soroos, Marvin S. (2002). 'Negotiating Our Climate', in Sharon L. Spray and Karen L. McGlothlin (eds.) *Global Climate Change*, Lanham: Rowman & Littlefield Publishers.

Skjærseth, Jon Birger and Tora Skodvin (2003). *Climate Change and the Oil Industry: Common Problem, Varying Strategies*, Manchester: Manchester University Press.

Skolnikoff, Eugene B. (1997). 'Same Science, Differing Policies: The Saga of Global Climate Change', MIT Joint Program on the Science and Policy of Global Change, 8, 22.

Smithey, Lee A. (2009). 'Social Movement Strategy, Tactics, and Collective Identity', *Sociology Compass*, 3(4): 658–71.

Spencer, Matthew (2010). 'Greenest Spending Review Ever?', [internet site]. Available: <http://www.green-alliance.org.uk/grea1.aspx?id=5219> Accessed 5 September 2012.

Sprinz, Detlef F. and Martin Weiß (2001). 'Domestic Politics and Global Climate Policy', in Urs Luterbacher and Detlef F. Sprinz (eds.) *International Relations and Global Climate Change*, Cambridge, Mass: MIT Press, pp. 67–94.

Stewart, John (2010). *Victory Against All the Odds*, Nottingham: Russell Press Ltd.

Stewart, John (2012). 'Heathrow's Third Runway Is Not Happening – Move On', *The Guardian*, 27 March, [internet site]. Available: <http://www.guardian.co.uk/commentisfree/2012/mar/27/heathrow-third-runway-not-happening> Accessed 07 May 2012.

Stoddard, Katy (2010). 'General Election 2010: Safe and Marginal Seats', *The Guardian*, 7 April, [internet site]. Available: <http://www.theguardian.com/news/datablog/2010/apr/07/election-safe-seats-electoral-reform> Accessed 16 August 2014.

Stratton, Allegra and Tim Webb (2010). 'Huhne Backtracks on Bank for Green Projects', *The Guardian*, 14 December, [internet site]. Available: <http://www.guardian.co.uk/politics/2010/dec/14/huhne-backtracks-bank-green-projects> Accessed 16 August 2012.

Strucke, James (2009). 'Live Coverage: Heathrow's Third Runway Decision', *The Guardian*, 15 January, [internet site]. Available: <http://www.guardian.co.uk/news/blog/2009/jan/15/heathrow-third-runway-travelandtransport> Accessed 7 May 2012.

Tarrow, Sidney (2011). *Power in Movement: Social Movements and Contentious Politics*, Cambridge: Cambridge University Press.

Taylor, Matthew (2010). 'Greenpeace Plans to Build Fortress on Heathrow Runway Site', *The Guardian*, 28 January, [internet site]. Available: <http://www.guardian.co.uk/environment/2010/jan/28/heathrow-third-runway-greenpeace> Accessed 07 May 2012.

Thatcher, Margaret (1988). 'Speech to the Royal Society', Fishmongers' Hall, 27 September: London <http://www.margaretthatcher.org/document/107346> accessed 22 June 2011.

Thatcher, Margaret (1989). 'Speech to United Nations General Assembly', United Nations Building, 8 November: New York. <http://www.margaretthatcher.org/document/107817> accessed 22 June 2011.

Thomas, Chris D., Alison Cameron, Rhys E. Green, Michel Bakkenes, Linda J. Beaumont, Yvonne C. Collingham, Barend F. N. Erasmus, Marinez Ferriera de Siqueira, Alan Grainger, Lee Hannah, Lesley Hughes, Brian Huntley, Albert S. Van Jaarsveld, Guy F. Midgley, Lera Miles, Miguel Ortega-Huerta, A. Townsend Peterson, Oliver L. Phillips, and Stephen E. Williams (2004). 'Extinction Risk from Climate Change', *Nature*, 427: 145–48.

Tipton, Jessica E. (2008). 'Why Did Russia Ratify the Kyoto Protocol? Why the Wait? An Analysis of the Environmental, Economic, and Political Debates', *Slovo*, 20(2): 67–96.

Toynbee, Polly (2008). 'This Craven Airport Decision Hands Cameron a Green Halo', *The Guardian*, 17 January [internet site]. Available: <http://www.theguardian.com/commentisfree/2009/jan/17/heathrow-runway> Accessed 17 August 2014.

Transform UK (2010). 'Businesses Demand £4 to 6 Billion for Green Investment Bank', [internet site]. Available: <http://www.transformuk.org/en/articles/1/businesses-demand-4-to-6-billion-for-green-investment-bank/> Accessed 17 April 2013.

Transform UK (2011a). 'Treasury Trolls Undermine Green Investment Bank', [internet site]. Available: <http://www.transformuk.org/en/articles/927/treasury-trolls-undermine-green-investment-bank/> Accessed 17 April 2013.

Transform UK (2011b). 'Treasury Clips Wings of Green Investment Bank', [internet site]. Available: <http://www.transformuk.org/en/articles/926/treasury-clips-wings-of-green-investment-bank/> Accessed 17 April 2013.

Transform UK (2011c). 'Green Investment Bank Master Plan Launched', [internet site]. Available: <http://www.transformuk.org/en/articles/929/green-investment-bank-master-plan-launched/> Accessed 17 April 2013.

Transform UK (2011d). *Clegg Announces Legislation to Set Up the Green Investment Bank and Full, Independent Borrowing Powers*, [internet site]. Available: <http://www.transformuk.org/en/articles/928/clegg-announces-legislation-to-set-up-the-green-investment-b/> Accessed 17 April 2013.

Transform UK and Client Earth (2012). *Priority Amendments to the Enterprise and Regulatory Reform Bill 2012 establishing the Green Investment Bank*, Memorandum to Public Bill Committee, [internet site]. Available: <http://www.publications.parliament.uk/pa/cm201213/cmpublic/enterprise/memo/err04.htm> Accessed 5 September 2012.

Tsutsui, Kiyoteru and Hwa J. Shin (2008). 'Global Norms, Local Activism, and Social Movement Outcomes: Global Human Rights and Resident Koreans in Japan', *Social Problems*, 55, 391–418.

UK Polling Report (2006a). 'Populus on the environment', 7 October [internet site]. Available: <http://ukpollingreport.co.uk/blog/archives/876> Accessed 28 April 2014.

UK Polling Report (2006b). 'Will the Dave the Chameleon Ad Backfire?', 23 April [internet site]. Available: <http://ukpollingreport.co.uk/blog/archives/193> Accessed 28 April 2014.

Van Rooy, Alison (1997). 'The Frontiers of Influence: NGO Lobbying at the 1974 World Food Conference, the 1992 Earth Summit and Beyond', *World Development*, 25(1): 93–114.

Vidal, John (2008). 'Thousands Expected at Carnival to Fight Heathrow Expansion', *The Guardian*, 31 May, [internet site]. Available: <http://www.guardian.co.uk/environment/2008/may/31/travelandtransport.communities> Accessed 28 April 2012.

Vidal, John and Tim Webb (2010). 'How Has DECC Fared in the Spending Review?', *The Guardian*, 20 October, [internet site]. Available: <http://www.guardian.co.uk/environment/2010/oct/20/decc-spending-review> Accessed 16 August 2012.

Walgrave, Stefaan, Stuart Soroka, and Michiel Nuytemans (2008). 'The Mass Media's Political Agenda-Setting Power: A Longitudinal Analysis of Media, Parliament, and Government in Belgium (1993 to 2000)', *Comparative Political Studies*, 41(6): 814–36.

Walters, Joanna (2002). 'BA to Buy Off Heathrow Locals', *The Observer*, 25 August, [internet site]. Available: <http://www.guardian.co.uk/business/2002/aug/25/transportintheuk.theobserver1> Accessed 28 April 2012.

Walsh, Edward J. (1981). 'Resource Mobilization and Citizen Protest in Communities around Three Mile Island', *Social Problems*, 29(1): 1–21.

Walsh, Willie (2008). 'Be Realistic about Heathrow Expansion', *The Guardian*, 15 December, [internet site]. Available: <http://www.guardian.co.uk/commentisfree/2008/dec/15/theairlineindustry-heathrow> Accessed 28 April 2012.

Weart, Spencer R. (2004). *The Discovery of Global Warming*, Cambridge: Harvard University Press.

Weart, Spencer (2011). 'The Oxford Handbook of Climate Change and Society', in John S. Dryzek, Richard B. Norgaard, and David Schlosberg (eds.) *The Oxford Handbook of Climate Change and Society*, Oxford: Oxford University Press, pp. 67–81.

Webb, Tim and Damian Carrington (2010). 'Chris Huhne Signals Frustration with Treasury over Green Investment Bank', *The Guardian*, 18 November, [internet site]. Available: <http://www.guardian.co.uk/environment/2010/nov/18/chris-huhne-green-investment-bank> Accessed 16 August 2012.

Webb, Keith, Ekkart Zimmermann, Michael Marsh, Annie-Marie Aish, Christina Mironseco, Christopher Mitchell, Leonardo Morlino, and James Walston (1983). 'Etiology and Outcomes of Protest', *American Behavioral Scientist*, 26(3): 311–31.

White, Michael and Anne Perkins (2002). '"Nasty Party" Warning to Tories', *The Guardian*, 8 October [internet site]. Available: <http://www.theguardian.com/politics/2002/oct/08/uk.conservatives2002> Accessed 22 December 2013.

Willetts, Peter (1996). 'From Stockholm to Rio and Beyond: The Impact of the Environmental Movement On the United Nations Consultative Arrangements for NGOs', *Review of International Studies*, 22(1): 57–80.

Wood, B. Dan and Jeffrey S. Peake (1998). 'The Dynamics of Foreign Policy Agenda Setting', *The American Political Science Review*, 92(1): 173–84.

Worthington, Bryony (2011). 'Bryony Worthington Speaking at the CDKN Action Lab', [internet site]. Available: <http://www.youtube.com/watch?v=X3xseCcfMZY> Accessed 22 December 2012.

YouGov (2006). 'YouGov / Sunday Times Survey Results', [internet site]. Available: <http://www.yougov.co.uk/archives/pdf/STI060101007_1.pdf> Accessed 29 March 2012.

YouGov (2007a). 'YouGov / Sunday Times Survey Results', [internet site]. Available: <http://cdn.yougov.com/today_uk_import/YG-Archives-pol-stimes-vi-070514.pdf> Accessed 29 March 2012.

YouGov (2007b). 'YouGov / Sky News Survey Results', [internet site]. Available: <http://cdn.yougov.com/today_uk_import/YG-Archives-pol-skynews-Brown Cameron-070627.pdf> Accessed 29 March 2012.

YouGov (2007c). 'YouGov / Sunday Times Survey', [internet site]. Available: <http://cdn.yougov.com/today_uk_import/YG-Archives-pol-stimes-vi-071008.pdf> Accessed 29 March 2012.

YouGov (2007d). 'YouGov / Daily Telegraph Survey Results', [internet site]. Available: <http://cdn.yougov.com/today_uk_import/YG-Archives-pol-dTel-vi-071026.pdf> Accessed 29 March 2012.

YouGov (2007e). 'YouGov / Sunday Times survey results', [internet site]. Available: <http://cdn.yougov.com/today_uk_import/YG-Archives-pol-stimes-endofyear-080108.pdf> Accessed 29 March 2012.

YouGov (2009). 'YouGov / The Sun Survey Results', [internet site]. Available: <http://cdn.yougov.com/today_uk_import/YG-Archives-pol-sun-vi-090109.pdf> Accessed 29 March 2012.

YouGov and The Taxpayers Alliance (2011). 'YouGov / The Taxpayers Alliance Survey Results', [internet site]. Available: <http://www.taxpayersalliance.com/spendingpoll2011.pdf> Accessed 30 June 2013.

Youth Sourcebook on Sustainable Development (1995). Winnipeg: IISD, [internet site]. Available: <http://iisd.ca/youth/ysbk000.htm> Accessed 30 June 2014.

Zillman, John W. (2009). 'A History of Climate Activities', *WMO Bulletin*, 58(3): 141–50.

Index

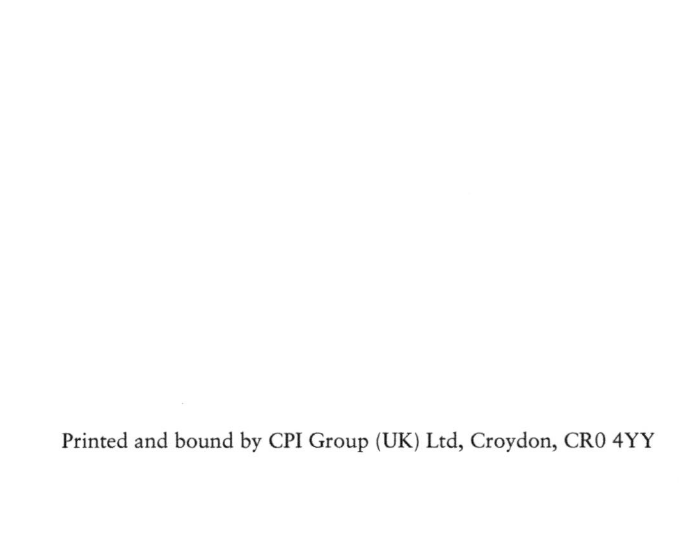

Printed and bound by CPI Group (UK) Ltd, Croydon, CR0 4YY